Elements of Numerical Methods for Compressible Flows

The purpose of this book is to present the basic elements of numerical methods for compressible flows. It is suitable for an advanced undergraduate or graduate course, and for specialists working in high-speed flows. The book focuses on the unsteady one-dimensional Euler equations, which form the basis for development of numerical algorithms in compressible fluid mechanics. The book is restricted to the basic concepts of finite volume methods, and is intended to provide the foundation for further study and application by the reader. The text is supplemented by end-of-chapter exercises.

Doyle D. Knight is Professor of Aerospace and Mechanical Engineering in the Department of Mechanical and Aerospace Engineering at Rutgers – The State University of New Jersey. He received his PhD in Aeronautics at the California Institute of Technology in 1974. After two years service in the United States Air Force as an Aeronautical Engineer and a one-year postdoctoral fellowship at the California Institute of Technology, he joined the faculty of the Department of Mechanical and Aerospace Engineering at Rutgers in 1977. He is a member of the American Institute of Aeronautics and Astronautics, the American Society of Mechanical Engineers, and the European Mechanics Society.

Cambridge Aerospace Series

Editors
Wei Shyy
and
Michael J. Rycroft

1. J. M. Rolfe and K. J. Staples (eds.): *Flight Simulation*
2. P. Berlin: *The Geostationary Applications Satellite*
3. M. J. T. Smith: *Aircraft Noise*
4. N. X. Vinh: *Flight Mechanics of High-Performance Aircraft*
5. W. A. Mair and D. L. Birdsall: *Aircraft Performance*
6. M. J. Abzug and E. E. Larrabee: *Airplane Stability and Control*
7. M. J. Sidi: *Spacecraft Dynamics and Control*
8. J. D. Anderson: *A History of Aerodynamics*
9. A. M. Cruise, J. A. Bowles, C. V. Goodall, and T. J. Patrick: *Principles of Space Instrument Design*
10. G. A. Khoury and J. D. Gillett (eds.): *Airship Technology*
11. J. Fielding: *Introduction to Aircraft Design*
12. J. G. Leishman: *Principles of Helicopter Aerodynamics*, second edition
13. J. Katz and A. Plotkin: *Low Speed Aerodynamics*, second edition
14. M. J. Abzug and E. E. Larrabee: *Airplane Stability and Control: A History of the Technologies that Made Aviation Possible*, second edition
15. D. H. Hodges and G. A. Pierce: *Introduction to Structural Dynamics and Aeroelasticity*
16. W. Fehse: *Automated Rendezvous and Docking of Spacecraft*
17. R. D. Flack: *Fundamentals of Jet Propulsion with Applications*
18. E. A. Baskharone: *Principles of Turbomachinery in Air-Breathing Engines*
19. Doyle D. Knight: *Numerical Methods for Compressible Flows*

Elements of Numerical Methods for Compressible Flows

DOYLE D. KNIGHT

Rutgers – The State University of New Jersey

CAMBRIDGE
UNIVERSITY PRESS

32 Avenue of the Americas, New York NY 10013-2473, USA

Cambridge University Press is part of the University of Cambridge.

It furthers the University's mission by disseminating knowledge in the pursuit of education, learning and research at the highest international levels of excellence.

www.cambridge.org
Information on this title: www.cambridge.org/9780521554749

© Cambridge University Press 2006

First published 2006
First paperback edition 2012

A catalogue record for this publication is available from the British Library

ISBN 978-0-521-55474-9 Hardback
ISBN 978-1-107-40702-2 Paperback

To Kelly, Courtney, and Caitlin

Contents

List of Illustrations

List of Tables

Preface

The purpose of this book is to present the basic elements of numerical methods for compressible flows. The focus is on the unsteady one-dimensional Euler equations which form the basis for numerical algorithms in compressible fluid mechanics. The book is restricted to the basic concepts of finite volume methods, and even in this regards is not intended to be exhaustive in its treatment. Several noteworthy texts on numerical methods for compressible flows are cited herein.

I would like to express my appreciation to Florence Padgett and Peter Gordon (Cambridge University Press) and Robert Stengel (Princeton University) for their patience. Any omissions or errors are mine alone.

New Brunswick, NJ

1
Governing Equations

All analyses concerning the motion of compressible fluids must necessarily begin, either directly or indirectly, with the statements of the four basic physical laws governing such motions.

A. Shapiro (1953)

1.1 Introduction

This chapter presents the governing equations of one-dimensional unsteady flow of a compressible fluid without derivation.† The following assumptions are made. First, the fluid is assumed to be calorically perfect, *i.e.*, the specific heats c_v and c_p at constant volume and pressure, respectively, are constant. Thus, the internal energy per unit mass e_i is

$$e_i = c_v T \tag{1.1}$$

where T is the static temperature. The static enthalpy h is defined as

$$h = e_i + \frac{p}{\rho} \tag{1.2}$$

Second, the fluid is assumed to be thermally perfect,

$$p = \rho R T \tag{1.3}$$

where p is the static pressure, ρ is the density, and R is the species gas constant,

$$R = c_p - c_v \tag{1.4}$$

† The full equations of three-dimensional compressible viscous flow are presented, for example, in Schreier (1982) and White (1974) .

1

It follows that

$$h = c_p T \tag{1.5}$$

Third, the fluid is assumed to be inviscid. Fourth, radiation effects and chemical reactions are omitted, and the fluid is assumed to be homogeneous (*i.e.*, uniform molecular composition).

1.2 Conservation Laws

Consider one-dimensional, inviscid unsteady flow in a tube of constant cross-sectional area A. We define a control volume V as shown in Fig. 1.1.

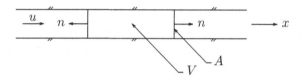

Fig. 1.1. Control volume

The integral conservation equations for mass, momentum, and energy are

$$\frac{d}{dt} \int_V \rho \, dV + \int_A \rho u n \, dA = 0 \tag{1.6}$$

$$\frac{d}{dt} \int_V \rho u \, dV + \int_A \left(\rho u^2 + p \right) n \, dA = 0 \tag{1.7}$$

$$\frac{d}{dt} \int_V \rho e \, dV + \int_A \left(\rho e + p \right) u n \, dA = 0 \tag{1.8}$$

where n is the unit vector in the outward direction† and the total energy e is

$$e = e_i + \tfrac{1}{2} u^2 \tag{1.9}$$

The differential forms of the conservation laws are

$$\frac{\partial \rho}{\partial t} + \frac{\partial \rho u}{\partial x} = 0 \tag{1.10}$$

$$\frac{\partial \rho u}{\partial t} + \frac{\partial \rho u^2}{\partial x} = -\frac{\partial p}{\partial x} \tag{1.11}$$

$$\frac{\partial \rho e}{\partial t} + \frac{\partial \left(\rho e + p \right) u}{\partial x} = 0 \tag{1.12}$$

† On the left face $n = -1$, and on the right face $n = +1$.

1.3 Convective Derivative

The convective derivative of a function f is defined as

$$\frac{Df}{Dt} = \frac{\partial f}{\partial t} + u\frac{\partial f}{\partial x} \tag{1.13}$$

and represents the rate of change of the variable f with respect to time while following a fluid particle. The differential form of the conservation laws for momentum and energy can be rewritten in terms of the convective derivative using the conservation of mass,

$$\frac{\partial \rho}{\partial t} + \frac{\partial \rho u}{\partial x} = 0 \tag{1.14}$$

to obtain

$$\rho\left(\frac{\partial u}{\partial t} + u\frac{\partial u}{\partial x}\right) = -\frac{\partial p}{\partial x} \tag{1.15}$$

$$\rho\left(\frac{\partial e}{\partial t} + u\frac{\partial e}{\partial x}\right) = -\frac{\partial pu}{\partial x} \tag{1.16}$$

1.4 Vector Notation

The differential form of the conservation laws (1.10) to (1.12) may be written in a compact vector notation as

$$\frac{\partial \mathcal{Q}}{\partial t} + \frac{\partial \mathcal{F}}{\partial x} = 0 \tag{1.17}$$

where

$$\mathcal{Q} = \left\{ \begin{array}{c} \rho \\ \rho u \\ \rho e \end{array} \right\} \tag{1.18}$$

$$\mathcal{F} = \left\{ \begin{array}{c} \rho u \\ \rho u^2 + p \\ \rho e u + p u \end{array} \right\} \tag{1.19}$$

1.5 Entropy

The change in entropy per unit mass s is

$$T ds = de_i - \frac{p}{\rho^2} d\rho \tag{1.20}$$

For a thermally perfect gas, this equation may be integrated to obtain

$$s - s_1 = c_v \ln \frac{T}{T_1} - R \ln \frac{\rho}{\rho_1} \tag{1.21}$$

Using (1.3), two alternate forms may be obtained:

$$s - s_1 = c_p \ln \frac{T}{T_1} - R \ln \frac{p}{p_1} \tag{1.22}$$

$$s - s_1 = c_v \ln \frac{p}{p_1} - c_p \ln \frac{\rho}{\rho_1} \tag{1.23}$$

Therefore, for isentropic flow between two states,

$$\frac{p_2}{p_1} = \left(\frac{\rho_2}{\rho_1}\right)^\gamma \tag{1.24}$$

$$\frac{p_2}{p_1} = \left(\frac{T_2}{T_1}\right)^{\gamma/(\gamma-1)} \tag{1.25}$$

The Second Law of Thermodynamics (Shapiro, 1953) may be expressed as

$$dS \geq \frac{\delta Q}{T} \tag{1.26}$$

where S is the entropy of a system (*i.e.*, an identifiable mass of fluid), δQ is the heat added to the system, and T is the static temperature. In particular, this implies that $dS \geq 0$ for an adiabatic process.

1.6 Speed of Sound

The speed of sound is the velocity of propagation of an infinitesimal disturbance in a quiescent fluid and is defined by

$$a = \sqrt{\left.\frac{\partial p}{\partial \rho}\right|_s} \tag{1.27}$$

where the partial derivative is taken at constant entropy s. For an ideal gas,

$$a = \sqrt{\gamma R T} \tag{1.28}$$

The Mach number is the ratio of the flow velocity to the speed of sound:

$$M = \frac{|u|}{a} \tag{1.29}$$

1.7 Alternate Forms

The total enthalpy H is

$$H = e + \frac{p}{\rho} \tag{1.30}$$

The conservation of energy may be expressed as an equation for the total enthalpy:

$$\frac{\partial \rho H}{\partial t} + \frac{\partial \rho H u}{\partial x} = \frac{\partial p}{\partial t} \tag{1.31}$$

Alternately, the energy and momentum equations may be utilized to obtain an equation for the internal energy e_i,

$$\frac{\partial \rho e_i}{\partial t} + \frac{\partial \rho e_i u}{\partial x} = -p\frac{\partial u}{\partial x} \tag{1.32}$$

Also, the energy may be rewritten in terms of the entropy. Using (1.20),

$$\frac{\partial \rho s}{\partial t} + \frac{\partial \rho s u}{\partial x} = 0 \tag{1.33}$$

It is noted that equations (1.31) to (1.33) can be rewritten in terms of the convective derivative using (1.10):

$$\rho\left(\frac{\partial H}{\partial t} + u\frac{\partial H}{\partial x}\right) = \frac{\partial p}{\partial t} \tag{1.34}$$

$$\rho\left(\frac{\partial e_i}{\partial t} + u\frac{\partial e_i}{\partial x}\right) = -p\frac{\partial u}{\partial x} \tag{1.35}$$

$$\rho\left(\frac{\partial s}{\partial t} + u\frac{\partial s}{\partial x}\right) = 0 \tag{1.36}$$

Exercises

1.1 Derive the alternate form of the momentum equation (1.15) using (1.11) and (1.10).

SOLUTION

The difference between (1.11) and (1.15) is the expression on the left side:

$$
\begin{aligned}
\frac{\partial \rho u}{\partial t} + \frac{\partial \rho u^2}{\partial x} &= \rho\frac{\partial u}{\partial t} + u\frac{\partial \rho}{\partial t} + \rho u\frac{\partial u}{\partial x} + u\frac{\partial \rho u}{\partial x} \\
&= \rho\left(\frac{\partial u}{\partial t} + u\frac{\partial u}{\partial x}\right) + u\left(\frac{\partial \rho}{\partial t} + \frac{\partial \rho u}{\partial x}\right) \\
&= \rho\left(\frac{\partial u}{\partial t} + u\frac{\partial u}{\partial x}\right)
\end{aligned}
$$

using (1.10). A similar derivation applies to (1.16).

1.2   Derive the enthalpy equation (1.31).

1.3   Derive the internal energy equation (1.32).

SOLUTION

Multiply the momentum equation (1.15) by u:

$$\rho u \left(\frac{\partial u}{\partial t} + u \frac{\partial u}{\partial x} \right) = -u \frac{\partial p}{\partial x}$$

$$\rho \left[\frac{\partial}{\partial t} \left(\tfrac{1}{2} u^2 \right) + u \frac{\partial}{\partial x} \left(\tfrac{1}{2} u^2 \right) \right] = -u \frac{\partial p}{\partial x}$$

which represents an equation for the kinetic energy† per unit mass $\tfrac{1}{2} u^2$. Subtract from the energy equation (1.16) using (1.9) to yield

$$\rho \left(\frac{\partial e_i}{\partial t} + u \frac{\partial e_i}{\partial x} \right) = \frac{\partial}{\partial x} (-pu) + u \frac{\partial p}{\partial x}$$

$$= -u \frac{\partial p}{\partial x} - p \frac{\partial u}{\partial x} + u \frac{\partial p}{\partial x}$$

$$= -p \frac{\partial u}{\partial x}$$

Multiply the mass equation (1.10) by e_i and add to the above to yield

$$\frac{\partial \rho e_i}{\partial t} + \frac{\partial \rho e_i u}{\partial x} = -p \frac{\partial u}{\partial x}$$

1.4   Derive the entropy equation (1.33).

1.5   Derive a conservation equation for the static enthalpy h.

SOLUTION

Add $\partial p / \partial t$ to both sides of the energy equation (1.12) to yield

$$\frac{\partial \rho H}{\partial t} + \frac{\partial \rho H u}{\partial x} = \frac{\partial p}{\partial t}$$

using (1.30). Multiply the mass equation by $\tfrac{1}{2} u^2$ and add to the kinetic energy equation (see above) to yield

$$\frac{\partial}{\partial t} \left(\tfrac{1}{2} \rho u^2 \right) + \frac{\partial}{\partial x} \left(\tfrac{1}{2} \rho u^2 u \right) = -u \frac{\partial p}{\partial x}$$

Subtract from the previous equation to yield

$$\frac{\partial \rho h}{\partial t} + \frac{\partial \rho h u}{\partial x} = \frac{\partial p}{\partial t} + u \frac{\partial p}{\partial x}$$

Using the mass equation, this may also be written as

$$\rho \frac{Dh}{Dt} = \frac{Dp}{Dt}$$

† The equation is also known as the *mechanical energy equation*.

1.6   In the presence of a body force per unit mass f, the momentum and energy equations become

$$\frac{\partial \rho u}{\partial t} + \frac{\partial \rho u^2}{\partial x} = -\frac{\partial p}{\partial x} + \rho f$$

$$\frac{\partial \rho e}{\partial t} + \frac{\partial (\rho e + p)\, u}{\partial x} = \rho f u$$

The mass equation is unchanged. Show that the total enthalpy equation is

$$\frac{\partial \rho H}{\partial t} + \frac{\partial \rho H u}{\partial x} = \frac{\partial p}{\partial t} + \rho f u$$

1.7   Derive the following equation:

$$\frac{1}{p}\frac{Dp}{Dt} = \frac{1}{c_v}\frac{Ds}{Dt} + \frac{\gamma}{\rho}\frac{D\rho}{Dt}$$

and provide a physical interpretation of each of the terms.

SOLUTION

From the entropy equation (1.23),

$$s - s_1 = c_v \ln \frac{p}{p_1} - c_p \ln \frac{\rho}{\rho_1}$$

Differentiating,

$$\frac{Ds}{Dt} = \frac{c_v}{p}\frac{Dp}{Dt} - \frac{c_p}{\rho}\frac{D\rho}{Dt}$$

Thus,

$$\frac{1}{p}\frac{Dp}{Dt} = \frac{1}{c_v}\frac{Ds}{Dt} + \frac{\gamma}{\rho}\frac{D\rho}{Dt}$$

The term on the left is the normalized rate of change of the static pressure following a fluid particle. The first term on the right is the rate of change of entropy (divided by c_v) following a fluid particle. Thus, an increase in entropy of the fluid particle acts to increase its static pressure. For an inviscid, homogeneous flow, the rate of change of entropy following a fluid particle is zero from (1.36), except when the fluid particle crosses a discontinuity (*i.e.*, at locations where the derivatives of the flow variables are not defined). The second term on the right is the rate of change of the fluid particle density (divided by ρ/γ). Thus, an increase in fluid particle density acts to increase the static pressure.

1.8   The total pressure p_o at a point is defined as the static pressure achieved by bringing a fluid particle at that point to rest isentropically. Similarly, the total temperature T_o at a point is defined as the static temperature achieved by bringing a fluid particle at that point to rest adiabatically. Therefore,

$$p_o = p\left[1 + \frac{(\gamma - 1)}{2}M^2\right]^{\gamma/(\gamma-1)}$$

$$T_o = T\left[1 + \frac{(\gamma - 1)}{2}M^2\right]$$

Thus, the entropy definition (1.25) may be written

$$s - s_1 = c_p \ln \frac{T_o}{T_{o_1}} - R \ln \frac{p_o}{p_{o_1}}$$

Show that

$$\frac{\rho}{\rho_o} \frac{Dp_o}{Dt} = \frac{\partial p}{\partial t} - \rho T_o \frac{Ds}{Dt}$$

where $\rho_o = p_o / RT_o$. Provide a physical interpretation of each of the terms.

1.9   The mechanical energy equation (see Problem 1.3) is

$$\frac{D}{Dt} \left(\tfrac{1}{2} u^2 \right) = -\frac{u}{\rho} \frac{\partial p}{\partial x}$$

Provide a physical explanation.

SOLUTION

The left-hand side of the equation is the time rate-of-change of the kinetic energy per unit mass following a fluid particle. The right-hand side is the work done on the unit mass by the pressure gradient. If the pressure increases in the direction of the flow (*i.e.*, $u \partial p / \partial x > 0$, which implies either a) $u > 0$ and $\partial p / \partial x > 0$ or b) $u < 0$ and $\partial p / \partial x < 0$), the kinetic energy decreases because the pressure gradient decelerates the flow. If the pressure decreases in the direction of the flow (*i.e.*, $u \partial p / \partial x < 0$), the kinetic energy increases because the pressure gradient accelerates the flow.

1.10   Derive an equation for the convective derivative of the Gibbs free energy

$$g = h - Ts$$

2

One-Dimensional Euler Equations

Nature confronts the observer with a wealth of nonlinear wave phenomena, not only in the flow of compressible fluids, but also in many other cases of practical interest.

R. Courant and K. O. Friedrichs (1948)

2.1 Introduction

The remainder of this book focuses on numerical algorithms for the unsteady Euler equations in one dimension. Although the practical applications of the one-dimensional Euler equations are certainly limited *per se*, virtually all numerical algorithms for inviscid compressible flow in two and three dimensions owe their origin to techniques developed in the context of the one-dimensional Euler equations. It is therefore essential to understand the development and implementation of these algorithms in their original one-dimensional context.

This chapter describes the principal mathematical properties of the one-dimensional Euler equations. An understanding of these properties is essential to the development of numerical algorithms. The presentation herein is necessarily brief. For further details, the reader may consult, for example, Courant and Friedrichs (1948) and Landau and Lifshitz (1958).

2.2 Differential Forms of One-Dimensional Euler Equations

The one-dimensional Euler equations can be expressed in a variety of differential forms, of which three are particularly useful in the development of numerical algorithms. These forms are applicable where the flow variables are continuously differentiable. However, flow solutions may exhibit discontinuities that require separate treatment, as will be discussed later in Section 2.3.

2.2.1 Conservative Form

The one-dimensional Euler equations in conservative differential form are

$$\frac{\partial \mathcal{Q}}{\partial t} + \frac{\partial \mathcal{F}}{\partial x} = 0 \tag{2.1}$$

where $\mathcal{Q}(x, t)$ is the vector of dependent variables,

$$\mathcal{Q} = \left\{ \begin{array}{c} \mathcal{Q}_1 \\ \mathcal{Q}_2 \\ \mathcal{Q}_3 \end{array} \right\} = \left\{ \begin{array}{c} \rho \\ \rho u \\ \rho e \end{array} \right\} \tag{2.2}$$

where ρ is the density, u is the velocity component in the x-direction, and e is the total energy per unit mass. The flux vector $\mathcal{F}(x, t)$ is

$$\mathcal{F} = \left\{ \begin{array}{c} \mathcal{F}_1 \\ \mathcal{F}_2 \\ \mathcal{F}_3 \end{array} \right\} = \left\{ \begin{array}{c} \rho u \\ \rho u u + p \\ \rho e u + p u \end{array} \right\} \tag{2.3}$$

The static pressure p is obtained from

$$p = (\gamma - 1)\left(\rho e - \tfrac{1}{2}\rho u^2\right) \tag{2.4}$$

The flux vector $\mathcal{F}$ is a function of $\mathcal{Q}$:

$$\mathcal{F} = \left\{ \begin{array}{c} \mathcal{Q}_2 \\[2mm] \dfrac{\mathcal{Q}_2^2}{\mathcal{Q}_1} + (\gamma - 1)\left(\mathcal{Q}_3 - \tfrac{1}{2}\dfrac{\mathcal{Q}_2^2}{\mathcal{Q}_1}\right) \\[3mm] \dfrac{\mathcal{Q}_2 \mathcal{Q}_3}{\mathcal{Q}_1} + (\gamma - 1)\dfrac{\mathcal{Q}_2}{\mathcal{Q}_1}\left(\mathcal{Q}_3 - \tfrac{1}{2}\dfrac{\mathcal{Q}_2^2}{\mathcal{Q}_1}\right) \end{array} \right\} \tag{2.5}$$

The term *conservative form* arises from the observation that in a finite domain the mass, momentum, and energy are strictly conserved in this formulation. Consider a region $0 \leq x \leq L$. Integrating (2.1) over this region,

$$\int_0^L \left[\frac{\partial \mathcal{Q}}{\partial t} + \frac{\partial \mathcal{F}}{\partial x} \right] dx = 0$$

Thus

$$\frac{\partial}{\partial t} \int_0^L \mathcal{Q} dx = \mathcal{F}(0,t) - \mathcal{F}(L,t)$$

which indicates that the net increase in $\mathcal{Q}$ per unit time within the region is due to the net flux into the region. This is the correct statement of the physics, of course.

Many numerical methods are patterned after the conservative form (2.1). The principal motivation is the requirement that flows with discontinuities (*e.g.*, shock waves and contact surfaces) must be accurately simulated. Lax and Wendroff (1960), in their study of conservation laws, showed that numerical algorithms that are both conservative and convergent can approximate the exact solution, including discontinuities,† to an arbitrary precision depending on the fineness of the spatial and temporal meshes.

2.2.2 Nonconservative Form

Equation (2.1) can also be written as

$$\frac{\partial \mathcal{Q}}{\partial t} + \mathcal{A}\frac{\partial \mathcal{Q}}{\partial x} = 0 \tag{2.6}$$

where $\mathcal{A}$ is the *Jacobian matrix* defined by

$$\mathcal{A} = \frac{\partial \mathcal{F}}{\partial \mathcal{Q}} \tag{2.7}$$

and given by

$$\mathcal{A} = \left\{ \begin{matrix} 0 & 1 & 0 \\ \dfrac{(\gamma-3)}{2}\left(\dfrac{\mathcal{Q}_2}{\mathcal{Q}_1}\right)^2 & (3-\gamma)\dfrac{\mathcal{Q}_2}{\mathcal{Q}_1} & (\gamma-1) \\ -\gamma\dfrac{\mathcal{Q}_2\mathcal{Q}_3}{\mathcal{Q}_1^2} + (\gamma-1)\left(\dfrac{\mathcal{Q}_2}{\mathcal{Q}_1}\right)^3 & \gamma\dfrac{\mathcal{Q}_3}{\mathcal{Q}_1} - \dfrac{3}{2}(\gamma-1)\left(\dfrac{\mathcal{Q}_2}{\mathcal{Q}_1}\right)^2 & \gamma\dfrac{\mathcal{Q}_2}{\mathcal{Q}_1} \end{matrix} \right\} \tag{2.8}$$

† The exact solution satisfies (2.1) almost everywhere, *i.e.*, at all points x in a finite domain except for a finite number of discontinuities where the flow variables satisfy the jump conditions discussed in Section 2.3.

Using the definition of Q, the Jacobian may be written in terms of u and e,

$$\mathcal{A} = \left\{ \begin{matrix} 0 & 1 & 0 \\ \dfrac{(\gamma-3)}{2}u^2 & (3-\gamma)u & (\gamma-1) \\ -\gamma eu + (\gamma-1)u^3 & \gamma e - \tfrac{3}{2}(\gamma-1)u^2 & \gamma u \end{matrix} \right\} \tag{2.9}$$

Alternately, using

$$e = \frac{a^2}{\gamma(\gamma-1)} + \tfrac{1}{2}u^2$$

the Jacobian may be written in terms of u and a,

$$\mathcal{A} = \left\{ \begin{matrix} 0 & 1 & 0 \\ \dfrac{(\gamma-3)}{2}u^2 & (3-\gamma)u & (\gamma-1) \\ -\dfrac{ua^2}{(\gamma-1)} + \dfrac{(\gamma-2)}{2}u^3 & \dfrac{a^2}{(\gamma-1)} + \dfrac{(3-2\gamma)}{2}u^2 & \gamma u \end{matrix} \right\} \tag{2.10}$$

Also, using the definition of the total enthalpy per unit mass H,

$$H = c_p T + \tfrac{1}{2}u^2$$

the Jacobian can be written in terms of u and H,

$$\mathcal{A} = \left\{ \begin{matrix} 0 & 1 & 0 \\ \dfrac{(\gamma-3)}{2}u^2 & (3-\gamma)u & (\gamma-1) \\ \dfrac{(\gamma-1)}{2}u^3 - Hu & H - (\gamma-1)u^2 & \gamma u \end{matrix} \right\} \tag{2.11}$$

There are a number of important properties of the Jacobian matrix $\mathcal{A}$. It is straightforward to show that its eigenvalues are

$$\begin{aligned} \lambda_1 &= u \\ \lambda_2 &= u + a \\ \lambda_3 &= u - a \end{aligned} \tag{2.12}$$

Equation (2.6) is *hyperbolic* since all of the eigenvalues of $\mathcal{A}$ are real (Garabedian, 1964). The corresponding right eigenvectors are

$$r_1 = \left\{ \begin{matrix} 1 \\ u \\ \tfrac{1}{2}u^2 \end{matrix} \right\}, \quad r_2 = \left\{ \begin{matrix} 1 \\ u + a \\ H + ua \end{matrix} \right\}, \quad r_3 = \left\{ \begin{matrix} 1 \\ u - a \\ H - ua \end{matrix} \right\} \tag{2.13}$$

and the corresponding left eigenvectors are

$$l_1 = \left\{ 1 - \frac{(\gamma - 1)}{2}\frac{u^2}{a^2}, \ (\gamma - 1)\frac{u}{a^2}, \ -\frac{(\gamma - 1)}{a^2} \right\}$$

$$l_2 = \left\{ \frac{(\gamma - 1)}{4}\frac{u^2}{a^2} - \frac{u}{2a}, \ -\frac{(\gamma - 1)}{2}\frac{u}{a^2} + \frac{1}{2a}, \ \frac{(\gamma - 1)}{2a^2} \right\}$$

$$l_3 = \left\{ \frac{(\gamma - 1)}{4}\frac{u^2}{a^2} + \frac{u}{2a}, \ -\frac{(\gamma - 1)}{2}\frac{u}{a^2} - \frac{1}{2a}, \ \frac{(\gamma - 1)}{2a^2} \right\} \qquad (2.14)$$

The Jacobian matrix $\mathcal{A}$ can be expressed as[†]

$$\mathcal{A} = T\Lambda T^{-1} \qquad (2.15)$$

where Λ is a diagonal matrix of eigenvalues of $\mathcal{A}$,

$$\Lambda = \left\{ \begin{matrix} \lambda_1 & 0 & 0 \\ 0 & \lambda_2 & 0 \\ 0 & 0 & \lambda_3 \end{matrix} \right\} \qquad (2.16)$$

The matrix[‡] T is the concatenation of the right eigenvectors of $\mathcal{A}$,

$$T = \left\{ \begin{matrix} 1 & 1 & 1 \\ u & u+a & u-a \\ \frac{1}{2}u^2 & H+ua & H-ua \end{matrix} \right\} \qquad (2.17)$$

The inverse T^{-1} is the concatenation of the left eigenvectors of $\mathcal{A}$,

$$T^{-1} = \left\{ \begin{matrix} 1 - \dfrac{(\gamma - 1)}{2}\dfrac{u^2}{a^2} & (\gamma - 1)\dfrac{u}{a^2} & -\dfrac{(\gamma - 1)}{a^2} \\[2ex] \dfrac{(\gamma - 1)}{4}\dfrac{u^2}{a^2} - \dfrac{1}{2}\dfrac{u}{a} & -\dfrac{(\gamma - 1)}{2}\dfrac{u}{a^2} + \dfrac{1}{2a} & \dfrac{(\gamma - 1)}{2a^2} \\[2ex] \dfrac{(\gamma - 1)}{4}\dfrac{u^2}{a^2} + \dfrac{1}{2}\dfrac{u}{a} & -\dfrac{(\gamma - 1)}{2}\dfrac{u}{a^2} - \dfrac{1}{2a} & \dfrac{(\gamma - 1)}{2a^2} \end{matrix} \right\} \qquad (2.18)$$

Two relationships between the left and right eigenvectors can be derived. Denote

$$r_k = \left\{ \begin{matrix} r_{k_1} \\ r_{k_2} \\ r_{k_3} \end{matrix} \right\} \qquad \text{for } k = 1, 2, 3 \qquad (2.19)$$

[†] Equation (2.15) indicates that $\mathcal{A}$ is similar to the diagonal matrix Λ. This is true since $\mathcal{A}$ has three linearly independent eigenvectors (Franklin, 1968).

[‡] The matrix T is not to be confused with the static temperature.

and

$$l_k = \left\{ \begin{array}{ccc} l_{k_1} & l_{k_2} & l_{k_3} \end{array} \right\} \quad \text{for } k = 1, 2, 3 \tag{2.20}$$

Then it may be directly verified†

$$\sum_{k=1}^{3} r_{i_k} l_{j_k} = \delta_{ij} \tag{2.21}$$

where δ_{ij} is the Kronecker delta defined by

$$\delta_{ij} = \left\{ \begin{array}{ll} 1 & \text{if } i = j \\ 0 & \text{otherwise} \end{array} \right. \tag{2.22}$$

The second relationship between the left and right eigenvectors can be found from the identity $TT^{-1} = I$. Since T is the concatenation of the eigenvectors r_k (taken as column vectors) and T^{-1} is the concatenation of the eigenvectors l_k (taken as row vectors), the identity $TT^{-1} = I$ yields

$$\sum_{k=1}^{3} r_{k_i} l_{k_j} = \delta_{ij} \tag{2.23}$$

It can also be shown by direct substitution (Exercise 2.5) that

$$\mathcal{F} = \mathcal{A}\mathcal{Q} \tag{2.24}$$

This is known as *Euler's Identity* or the *homogeneity property* and is the basis of some of the *Flux Vector Splitting* methods that will be discussed in Chapter 6.

2.2.3 Characteristic Form

The one-dimensional Euler equations may be written in a form that exhibits their *wavelike* character. It is first useful to describe wavelike behavior. Consider the scalar equation

$$\frac{\partial f}{\partial t} + c \frac{\partial f}{\partial x} = 0 \tag{2.25}$$

where $f(x, t)$ is a scalar function and c is a positive constant. For an unbounded domain $-\infty < x < \infty$ and initial condition $f(x, 0) = g(x)$, the solution to (2.25) is

$$f(x, t) = g(x - ct) \tag{2.26}$$

---

† The relationship (2.21) follows directly from the fact that the eigenvalues of $\mathcal{A}$ are distinct (Isaacson and Keller, 1966).

This represents a wave traveling to the right with velocity c without a change of form.

The wavelike character of (2.25) can also be discerned by considering the behavior of $f(x,t)$ along a curve $x(t)$ as shown in Fig. 2.1.

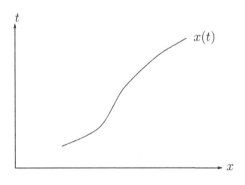

Fig. 2.1. Curve $x(t)$

Denoting the derivative of f with respect to t along the curve $x(t)$ by df/dt, we then have

$$\frac{df}{dt} = \frac{d}{dt}f(x,t) = \frac{\partial f}{\partial x}\frac{dx(t)}{dt} + \frac{\partial f}{\partial t} \tag{2.27}$$

where $\partial f/\partial x$ denotes the partial derivative of f with respect to x holding t constant, and likewise $\partial f/\partial t$ indicates the partial derivative of f with respect to t holding x constant. Comparing (2.25) and (2.27),

$$\frac{df}{dt} = 0 \quad \text{on} \quad \frac{dx}{dt} = c \tag{2.28}$$

which indicates that the value of f is unchanged on the curve $dx/dt = c$ or $x - ct = $ constant as seen in (2.26). The curves $x - ct = $ constant are known as the *characteristics* of (2.25).

The one-dimensional Euler equations can be recast in a form similar to (2.27). The equations are

$$\frac{\partial \rho}{\partial t} + \frac{\partial \rho u}{\partial x} = 0 \tag{2.29}$$

$$\frac{\partial \rho u}{\partial t} + \frac{\partial \rho u^2}{\partial x} = -\frac{\partial p}{\partial x} \tag{2.30}$$

$$\frac{\partial s}{\partial t} + u\frac{\partial s}{\partial x} = 0 \tag{2.31}$$

where the energy equation has been replaced by the equation for entropy

(1.36) assuming inviscid, nonheat conducting flow. It is evident that the entropy equation is already in the form of (2.28) and can be written as

$$\frac{ds}{dt} = 0 \quad \text{on} \quad \frac{dx(t)}{dt} = u(x(t), t) \tag{2.32}$$

which implies that the entropy is conserved following a fluid particle.

We seek a transformation of the remaining equations to a form similar to (2.27). From the definition of entropy (1.23),

$$s - s_1 = c_v \log\left(\frac{p}{p_1}\right) - c_p \log\left(\frac{\rho}{\rho_1}\right)$$

Then

$$\frac{\partial s}{\partial t} + u\frac{\partial s}{\partial x} = \frac{c_v}{p}\left(\frac{\partial p}{\partial t} + u\frac{\partial p}{\partial x}\right) - \frac{c_p}{\rho}\left(\frac{\partial \rho}{\partial t} + u\frac{\partial \rho}{\partial x}\right)$$

and thus

$$\frac{\partial p}{\partial t} + u\frac{\partial p}{\partial x} - \frac{\gamma p}{\rho}\left(\frac{\partial \rho}{\partial t} + u\frac{\partial \rho}{\partial x}\right) = 0$$

From the conservation of mass,

$$\frac{\partial \rho}{\partial t} + u\frac{\partial \rho}{\partial x} = -\rho\frac{\partial u}{\partial x}$$

and since $a^2 = \gamma p/\rho$,

$$\frac{\partial p}{\partial t} + u\frac{\partial p}{\partial x} + a^2\rho\frac{\partial u}{\partial x} = 0 \tag{2.33}$$

Using the conservation of mass, the conservation of momentum is rewritten as

$$\rho\frac{\partial u}{\partial t} + \rho u\frac{\partial u}{\partial x} = -\frac{\partial p}{\partial x}$$

and multiplying this equation by a and adding (2.33),

$$\frac{\partial p}{\partial t} + (u + a)\frac{\partial p}{\partial x} + \rho a\left[\frac{\partial u}{\partial t} + (u + a)\frac{\partial u}{\partial x}\right]$$

or

$$\frac{1}{\rho a}\frac{dp}{dt} + \frac{du}{dt} = 0 \quad \text{on} \quad \frac{dx}{dt} = u + a \tag{2.34}$$

Assuming that the entropy at some initial time is uniform, from (2.32) the entropy remains uniform, and thus ρ and a are functions of p alone. Hence, (2.34) may be rewritten as

$$\frac{d}{dt}\left[\int\frac{dp}{\rho a} + u\right] = 0$$

Since $p/p_1 = (\rho/\rho_1)^\gamma$ and $a/a_1 = (p/p_1)^{(\gamma-1)/2\gamma}$,

$$\frac{d}{dt}\left(\frac{2}{(\gamma-1)}a + u\right) = 0 \quad \text{on} \quad \frac{dx}{dt} = u + a$$

Similarly, it is possible to derive

$$\frac{d}{dt}\left(\frac{2}{(\gamma-1)}a - u\right) = 0 \quad \text{on} \quad \frac{dx}{dt} = u - a$$

The unsteady one-dimensional Euler equations may therefore be rewritten as

$$\frac{d}{dt}\left(\frac{2}{(\gamma-1)}a + u\right) = 0 \quad \text{on} \quad \frac{dx}{dt} = u + a$$

$$\frac{d}{dt}\left(\frac{2}{(\gamma-1)}a - u\right) = 0 \quad \text{on} \quad \frac{dx}{dt} = u - a$$

$$\frac{ds}{dt} = 0 \quad \text{on} \quad \frac{dx}{dt} = u \qquad (2.35)$$

These are known as *Riemann invariants*. Note that, in general, u and a are functions of x and t. The curves $dx/dt = u+a$, $dx/dt = u-a$, and $dx/dt = u$ are the *characteristics*. Equations (2.35) imply that the Riemann invariants are convected without change along their respective characteristics, *i.e.*, the Euler equations admit wavelike solutions. However, Equations (2.35) permit multiple-valued solutions that are unphysical and hence must be supplemented with additional constraints, *i.e.*, the solutions for discontinuous waves (Section 2.3).

2.3 Discontinuous Waves

The one-dimensional unsteady Euler equations admit waves wherein one or more flow variables are discontinuous across the wavefront. Consider a wavefront with velocity $u_w(t)$. As shown in Fig. 2.2, a small control volume δV spans the wavefront, where δV has width dx and height dy. Relative to an inertial reference frame, the fluid has velocities u_1 and u_2 to the left and right of the wavefront, respectively. The control volume is affixed to the wavefront and is therefore moving at velocity u_w relative to the reference frame.

The conservation of mass (1.6) applied to δV yields†

$$\frac{\partial \bar{\rho}}{\partial t}\,dx\,dy + \rho_2\,(u_2-u_w)\,dy - \rho_1\,(u_1-u_w)\,dy = 0$$

† Strictly, to the lowest order.

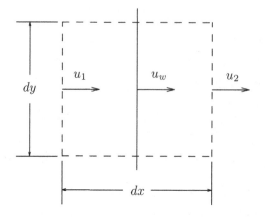

Fig. 2.2. Control volume affixed to wavefront

where $\bar{\rho} = \frac{1}{2}(\rho_1 + \rho_2)$. Dividing by dy and taking the limit $dx \to 0$,

$$\rho_1\,(u_1 - u_w) - \rho_2\,(u_2 - u_w) = 0$$

or

$$[\rho\,(u - u_w)]_w = 0 \qquad (2.36)$$

where we define

$$[f]_w = f_1 - f_2 \qquad (2.37)$$

where f_1 and f_2 represent the value of the function f on either side of the wave surface. Note that f_1 is not necessarily equal to f_2.

The conservation of x-momentum (1.7) yields

$$\frac{\partial \overline{\rho u}}{\partial t}\,dx\,dy + (\rho_2 u_2\,(u_2 - u_w) + p_2)\,dy - (\rho_1 u_1\,(u_1 - u_w) + p_1)\,dy = 0$$

and hence taking $dx \to 0$ yields

$$[\rho u\,(u - u_w) + p]_w = 0 \qquad (2.38)$$

The conservation of energy likewise yields

$$[\rho e\,(u - u_w) + p u]_w = 0 \qquad (2.39)$$

Equations (2.36) to (2.39) are the one-dimensional *Rankine-Hugoniot* conditions and are summarized in Table 2.1. They admit two different types of wavelike solutions.† The first is a *contact surface* defined by

$$u_1 = u_2 = u_w \quad \text{(contact surface)} \qquad (2.40)$$

† A third type of discontinuous solution (*vortex sheet*) is possible in two and three dimensions.

This solution satisfies (2.36). From (2.38), $p_1 = p_2$. Equation (2.39) is also satisfied. Note that there is no condition imposed on $[\rho]_w$ or $[\rho e]_w$. Therefore, it is possible that $\rho_1 \neq \rho_2$ and $\rho_1 e_1 \neq \rho_2 e_2$, with the latter implying that $T_1 \neq T_2$.

Table 2.1. *One-Dimensional Rankine-Hugoniot Conditions*

Equation	Condition
Mass	$[\rho(u - u_w)]_w = 0$
Momentum	$[\rho u(u - u_w) + p]_w = 0$
Energy	$[\rho e(u - u_w) + pu]_w = 0$

The second is a *shock wave* for which

$$u_1 \neq u_2 \neq u_w \quad \text{(shock wave)} \tag{2.41}$$

There are two cases, corresponding to the sign of $u_1 - u_w$. In the first case, $u_1 - u_w > 0$, which corresponds to a wave moving to the left (*i.e.*, the wave speed is negative relative to a frame of reference traveling at u_1). Equations (2.36) to (2.39) admit a one-parameter family of solutions for given u_1 and a_1, where the parameter may be chosen to be the shock pressure ratio[†] $\sigma_l = p_2/p_1$ that cannot be less than one.[‡] The result is[1]

$$\frac{\rho_2}{\rho_1} = \frac{(\gamma - 1) + (\gamma + 1)\sigma_l}{(\gamma + 1) + (\gamma - 1)\sigma_l} \tag{2.42}$$

$$\frac{T_2}{T_1} = \frac{\rho_1}{\rho_2}\sigma_l \tag{2.43}$$

$$u_2 = \frac{\rho_1}{\rho_2}\left[u_1 + u_w\left(\frac{\rho_2}{\rho_1} - 1\right)\right] \tag{2.44}$$

$$u_w = u_1 - a_1\left[\frac{(\gamma - 1)}{2\gamma} + \frac{(\gamma + 1)}{2\gamma}\sigma_l\right]^{1/2} \tag{2.45}$$

Note that Equation (2.44) can be rewritten as

$$u_2 = u_1 - \frac{a_1}{\gamma}\frac{(\sigma_l - 1)}{\sqrt{\frac{(\gamma+1)}{2\gamma}\sigma_l + \frac{(\gamma-1)}{2\gamma}}} \tag{2.46}$$

In the second case, $u_1 - u_w < 0$, which corresponds to a wave moving to the right. Equations (2.36) to (2.39) admit a one-parameter family of solutions

---

[†] The subscript l denotes a wave moving to the left.

[‡] The Second Law of Thermodynamics requires the entropy change to be nonnegative as the flow crosses the discontinuity. It can be shown that this implies $\sigma_l \geq 1$ (Liepmann and Roshko 1957).

for given u_2 and a_2, where the parameter may be chosen to be the shock pressure ratio $\sigma_r = p_1/p_2$, which cannot be less than one:

$$\frac{\rho_1}{\rho_2} = \frac{(\gamma - 1) + (\gamma + 1)\sigma_r}{(\gamma + 1) + (\gamma - 1)\sigma_r} \tag{2.47}$$

$$\frac{T_1}{T_2} = \frac{\rho_2}{\rho_1} \sigma_r \tag{2.48}$$

$$u_1 = \frac{\rho_2}{\rho_1} \left[u_2 + u_w \left(\frac{\rho_1}{\rho_2} - 1 \right) \right] \tag{2.49}$$

$$u_w = u_2 + a_2 \left[\frac{(\gamma - 1)}{2\gamma} + \frac{(\gamma + 1)}{2\gamma} \sigma_r \right]^{1/2} \tag{2.50}$$

Note that Equation (2.49) can be rewritten as

$$u_1 = u_2 + \frac{a_2}{\gamma} \frac{(\sigma_r - 1)}{\sqrt{\frac{(\gamma+1)}{2\gamma}\sigma_r + \frac{(\gamma-1)}{2\gamma}}} \tag{2.51}$$

2.4 Method of Characteristics

The characteristic form (2.35) of the one-dimensional Euler equations offers a direct method for solution. The equations are

$$\frac{d}{dt}\left(\frac{2}{(\gamma-1)} a + u \right) = 0 \qquad \text{on} \quad \frac{dx}{dt} = u + a \tag{2.52}$$

$$\frac{d}{dt}\left(\frac{2}{(\gamma-1)} a - u \right) = 0 \qquad \text{on} \quad \frac{dx}{dt} = u - a \tag{2.53}$$

$$\frac{ds}{dt} = 0 \qquad \text{on} \quad \frac{dx}{dt} = u \tag{2.54}$$

For simplicity, we consider an unbounded domain $-\infty < x < \infty$. The initial entropy is uniform and denoted by s_o. Initial continuous profiles of $u(x,t)$ and $a(x,t)$ are specified as

$$u(x,o) = u_o(x)$$
$$a(x,0) = a_o(x)$$

Consider a point† p at an infinitesimal time $dt > 0$ as shown in Fig. 2.3. There are three characteristics that emanate from $t = 0$ and intersect p:

$$\frac{dx}{dt} = u_1 + a_1$$

---

† The point p is not to be confused with the static pressure.

$$\frac{dx}{dt} = u_2 - a_2$$

$$\frac{dx}{dt} = u_3$$

where the slope of each characteristic is evaluated at its origin at $t = 0$.

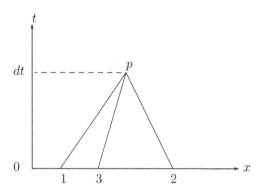

Fig. 2.3. The three characteristics intersecting at point p. The figure is drawn for $a > u > 0$.

From (2.52) and (2.53),

$$\frac{2}{(\gamma - 1)}a_p + u_p = \frac{2}{(\gamma - 1)}a_1 + u_1$$

$$\frac{2}{(\gamma - 1)}a_p - u_p = \frac{2}{(\gamma - 1)}a_2 - u_2$$

which may be solved to yield

$$a_p = \tfrac{1}{2}(a_1 + a_2) + \frac{(\gamma - 1)}{4}(u_1 - u_2) \tag{2.55}$$

$$u_p = \frac{1}{(\gamma - 1)}(a_1 - a_2) + \tfrac{1}{2}(u_1 + u_2) \tag{2.56}$$

From (2.54),

$$s_p = s_o \tag{2.57}$$

and thus the flow remains isentropic. Therefore,

$$p_p = p_r \left(\frac{a_p}{a_r}\right)^{2\gamma/(\gamma-1)} \tag{2.58}$$

$$\rho_p = \rho_r \left(\frac{a_p}{a_r}\right)^{2/(\gamma-1)} \tag{2.59}$$

where the subscript $_r$ indicates a reference value.

It would appear that equations (2.55) to (2.59) provide an algorithm for the complete solution to the flowfield at time dt, *i.e.*, these equations could be employed to determine the solution on a discrete set of points p. This process could presumably then be repeated and thus the solution could be obtained in principle. However, this is not the case. In fact, Equations (2.55) to (2.59) cannot be used solely in all cases because they do not guarantee that there will be *only* three characteristics intersecting point p, *i.e.*, they do not guarantee the uniqueness of the solution. Indeed, nonunique (*i.e.*, multiple-valued) solutions are mathematically possible in certain regions depending on the initial conditions and in the absence of any additional constraints. Since this is physically nonsensible, a modification is required which is the introduction of a discontinuous wave (*i.e.*, shock) when necessary. This issue will be discussed in further detail in Section 2.7.

2.5 Expansion Fan

The Method of Characteristics can be used to determine the flow solution within a simple wave joining two regions of uniform entropy at different velocities, pressures and temperatures. Figure 2.4 illustrates the first type of simple wave bounded by the positive characteristics $dx/dt = u_1 + a_1$ and $dx/dt = u_2 + a_2$ bordering two regions of uniform flow denoted by subscripts 1 and 2. The region between these two positive characteristics is denoted by an *expansion fan*. Consider points a and b on a positive characteristic $dx/dt = u + a$ located within the expansion fan. A negative characteristic $dx/dt = u - a$ drawn from a point c will intersect point b, and likewise a negative characteristic from point d will intersect point a. From (2.52),

$$\frac{2}{(\gamma - 1)} a_a + u_a = \frac{2}{(\gamma - 1)} a_b + u_b \tag{2.60}$$

From (2.53),

$$\frac{2}{(\gamma - 1)} a_a - u_a = \frac{2}{(\gamma - 1)} a_d - u_d \tag{2.61}$$

$$\frac{2}{(\gamma - 1)} a_b - u_b = \frac{2}{(\gamma - 1)} a_c - u_c \tag{2.62}$$

Since points c and d are in the uniform region 1,

$$\frac{2}{(\gamma - 1)} a_a - u_a = \frac{2}{(\gamma - 1)} a_b - u_b \tag{2.63}$$

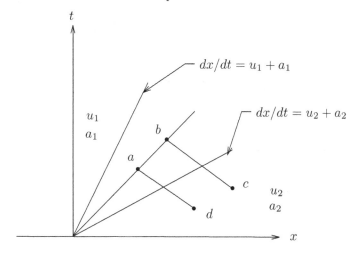

Fig. 2.4. Right-moving expansion wave

From (2.60) and (2.63),

$$u_a = u_b$$
$$a_a = a_b$$

and thus every characteristic $dx/dt = u + a$ within the expansion fan is a straight line. Now, from (2.61),

$$\frac{2}{(\gamma - 1)}a_a - u_a = \frac{2}{(\gamma - 1)}a_2 - u_2$$

and thus

$$a_a = \frac{(\gamma - 1)}{2}(u_a - u_2) + a_2$$

Therefore, the characteristic inside the expansion fan is

$$\frac{dx}{dt} = \frac{(\gamma + 1)}{2}u_a - \frac{(\gamma - 1)}{2}u_2 + a_2$$

Since u_a is constant on the characteristic, the equation can be integrated to obtain

$$x = \left[\frac{(\gamma + 1)}{2}u_a - \frac{(\gamma - 1)}{2}u_2 + a_2\right]t$$

and thus

$$u = \frac{2}{(\gamma + 1)}\left[\frac{x}{t} + \frac{(\gamma - 1)}{2}u_2 - a_2\right] \tag{2.64}$$

$$a = \frac{(\gamma - 1)}{(\gamma + 1)}\left[\frac{x}{t} - u_2\right] + \frac{2}{(\gamma + 1)}a_2 \tag{2.65}$$

which represents the solution for u and a within the expansion fan. The remaining flow variables can be obtained from the isentropic relations

$$\frac{p}{p_1} = \left(\frac{a}{a_1}\right)^{2\gamma/(\gamma-1)}$$

$$\frac{T}{T_1} = \left(\frac{a}{a_1}\right)^2$$

$$\frac{\rho}{\rho_1} = \left(\frac{a}{a_1}\right)^{2/(\gamma-1)} \tag{2.66}$$

In addition, by extending point a into region 1 and using Equation (2.61), the flow conditions in regions 1 and 2 are related by

$$u_1 = u_2 + \frac{2}{(\gamma-1)}(a_1 - a_2) \tag{2.67}$$

Using the isentropic relations, Equation (2.66), this becomes

$$u_1 = u_2 + \frac{2a_2}{(\gamma-1)}\left[\left(\frac{p_1}{p_2}\right)^{(\gamma-1)/2\gamma} - 1\right] \tag{2.68}$$

Figure 2.5 illustrates the second type of simple wave bounded by the negative characteristics $dx/dt = u_1 - a_1$ and $dx/dt = u_2 - a_2$ bordering two regions of uniform flow denoted by subscripts 1 and 2. An analysis similar to the first type yields the following solution within the expansion fan:

$$u = \frac{2}{(\gamma+1)}\left[\frac{x}{t} + \frac{(\gamma-1)}{2}u_1 + a_1\right] \tag{2.69}$$

$$a = \frac{(\gamma-1)}{(\gamma+1)}\left[-\frac{x}{t} + u_1\right] + \frac{2}{(\gamma+1)}a_1 \tag{2.70}$$

The remaining flow variables can be obtained from the isentropic relations in Equations (2.66). In addition, the flow conditions in regions 1 and 2 are related by

$$u_2 = u_1 + \frac{2}{(\gamma-1)}(a_1 - a_2) \tag{2.71}$$

Using the isentropic relations in Equation (2.66), this becomes

$$u_2 = u_1 - \frac{2a_1}{(\gamma-1)}\left[\left(\frac{p_2}{p_1}\right)^{(\gamma-1)/2\gamma} - 1\right] \tag{2.72}$$

It should be noted that, although Fig. 2.4 shows that both characteristics $dx/dt = u_1 + a_1$ and $dx/dt = u_2 + a_2$ move to the right, this need not be the case. An analogous conclusion holds for Fig. 2.5.

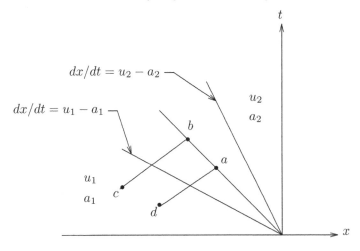

Fig. 2.5. Left-moving expansion wave

2.6 Domains of Dependence and Influence

It is evident from (2.55) to (2.58) that the flow at p depends only on a limited range of data at $t = 0$ in x, *i.e.*, the region

$$\min(x_1, x_2, x_3) \leq x \leq \max(x_1, x_2, x_3)$$

This region is known as the *domain of dependence* of the solution at point p, as illustrated in Fig. 2.6.

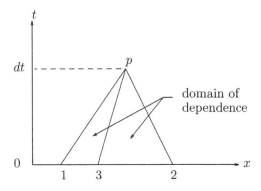

Fig. 2.6. The three characteristics intersecting at point p. The figure is drawn for $a > u > 0$.

Similarly, it is possible to define the points x_1, x_2, x_3, which are the intersection of the three characteristics emanating from a single point p at $t = 0$

with $t = dt$, as illustrated in Fig. 2.7. The region

$$\min(x_1, x_2, x_3) \leq x \leq \max(x_1, x_2, x_3)$$

at $t = dt$ defines the *domain of influence* of point p. A change of flow conditions at p at $t = 0$ will only influence the flow within this region at $t = dt$.

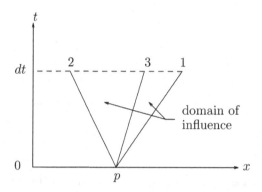

Fig. 2.7. The three characteristics emanating from point p. The figure is drawn for $a > u > 0$.

2.7 Shock Formation

Depending on the initial conditions, the formal application of the Method of Characteristics (Section 2.4) can yield nonunique solutions, *i.e.*, more than three characteristics intersect at some point p at some instant in time. This is physically impossible, and therefore one or more of the assumptions employed in deriving the characteristic equations (2.35) must be violated. Specifically, the assumption that the flowfield is continuous and differentiable is violated at discrete locations where a discontinuous solution of the Euler equations (*i.e.*, shock wave) appears.

The conditions for the formation of a shock wave and its position in space and time can be determined. We follow the derivation of Landau and Lifshitz (1959). Consider an adiabatic, inviscid flow with no shocks. The entropy s is assumed uniform at $t = 0$. The Euler equations are

$$\frac{\partial \rho}{\partial t} + \frac{\partial \rho u}{\partial x} = 0 \tag{2.73}$$

$$\frac{\partial \rho u}{\partial t} + \frac{\partial \rho u^2}{\partial x} = -\frac{\partial p}{\partial x} \tag{2.74}$$

$$\frac{\partial s}{\partial t} + u\frac{\partial s}{\partial x} = 0 \tag{2.75}$$

From the entropy equation, s remains constant for $t > 0$ until the possible formation of any shock waves.

We seek finite amplitude waves wherein the velocity u is a function of density ρ. The conservation of mass becomes

$$\frac{\partial \rho}{\partial t} + \frac{d\rho u}{d\rho}\frac{\partial \rho}{\partial x} = 0$$

and therefore

$$\frac{\partial \rho}{\partial t}\left(\frac{\partial \rho}{\partial x}\right)^{-1} = -\frac{d\rho u}{du} \tag{2.76}$$

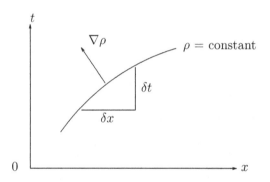

Fig. 2.8. Isocontours of density

Consider the isocontours of ρ in the $x - t$ plane (Fig. 2.8). The vector $\delta x \vec{e}_x + \delta t \vec{e}_t$ is constructed to be parallel to the isocontour, where $\vec{e}_x$ and $\vec{e}_t$ are unit vectors in the x and t directions, respectively. By definition, the gradient

$$\nabla \rho = \frac{\partial \rho}{\partial x}\vec{e}_x + \frac{\partial \rho}{\partial t}\vec{e}_t$$

is orthogonal to the isocontour. Therefore, the inner product of these vectors is zero:

$$\left(\frac{\partial \rho}{\partial x}\vec{e}_x + \frac{\partial \rho}{\partial t}\vec{e}_t\right) \cdot (\delta x \vec{e}_x + \delta t \vec{e}_t) = 0$$

from which

$$\frac{\partial \rho}{\partial t}\left(\frac{\partial \rho}{\partial x}\right)^{-1} = -\frac{\delta x}{\delta t} = -\left.\frac{\partial x}{\partial t}\right|_\rho \tag{2.77}$$

From (2.76) and (2.77),

$$\left.\frac{\partial x}{\partial t}\right|_{\rho} = \frac{d\rho u}{d\rho} \tag{2.78}$$

From the conservation of momentum, using the conservation of mass,

$$\frac{\partial u}{\partial t} + u\frac{\partial u}{\partial x} + \frac{1}{\rho}\frac{\partial p}{\partial x} = 0$$

Now

$$\frac{\partial p}{\partial x} = \left.\frac{\partial p}{\partial \rho}\right|_{s}\frac{\partial \rho}{\partial x} + \left.\frac{\partial p}{\partial s}\right|_{\rho}\frac{\partial s}{\partial x}$$

The second term is zero, and using (1.27) the first term is $a^2 \partial\rho/\partial x$, where a is the speed of sound. Since u is assumed to be a function of ρ,

$$\partial\rho/\partial x = \left(\frac{du}{d\rho}\right)^{-1}\frac{\partial u}{\partial x}$$

and the momentum equation can be rewritten as

$$\frac{\partial u}{\partial t}\left(\frac{\partial u}{\partial x}\right)^{-1} = -\left[u + \frac{a^2}{\rho}\left(\frac{du}{d\rho}\right)^{-1}\right] \tag{2.79}$$

By an argument similar to that leading to (2.77),

$$\frac{\partial u}{\partial t}\left(\frac{\partial u}{\partial x}\right)^{-1} = -\left.\frac{\partial x}{\partial t}\right|_{u} \tag{2.80}$$

and since u is a function of ρ,

$$\left.\frac{\partial x}{\partial t}\right|_{u} = \left.\frac{\partial x}{\partial t}\right|_{\rho}$$

From (2.78) and (2.79),

$$\frac{d\rho u}{d\rho} = u + \frac{a^2}{\rho}\left(\frac{du}{d\rho}\right)^{-1}$$

and thus

$$\frac{du}{d\rho} = \pm\frac{a}{\rho} \tag{2.81}$$

Therefore

$$u = \pm\int\frac{a}{\rho}d\rho \tag{2.82}$$

Since the flow is isentropic, using (1.24), (1.25), and (1.28),

$$u = \pm\frac{2}{(\gamma-1)}\int da$$

We assume that there is some point within the gas at $t = 0$ where the velocity is zero.† Then

$$u = \pm \frac{2}{(\gamma - 1)} (a - a_o)$$

where a_o is the speed of sound at the location where $u = 0$. Thus,

$$a = a_o \pm \frac{(\gamma - 1)}{2} u \qquad (2.83)$$

and therefore

$$\rho = \rho_o \left[1 \pm \frac{(\gamma - 1)}{2} \frac{u}{a_o} \right]^{2/(\gamma - 1)} \qquad (2.84)$$

$$p = p_o \left[1 \pm \frac{(\gamma - 1)}{2} \frac{u}{a_o} \right]^{2\gamma/(\gamma - 1)} \qquad (2.85)$$

where ρ_o and p_o are the density and pressure at the location where $u = 0$.

From (2.79), (2.80), and (2.81),

$$\left. \frac{\partial x}{\partial t} \right|_u = u \pm a \qquad (2.86)$$

and using (2.83),

$$\left. \frac{\partial x}{\partial t} \right|_u = \pm a_o + \frac{(\gamma + 1)}{2} u \qquad (2.87)$$

Therefore

$$x = \left[\pm a_o + \frac{(\gamma + 1)}{2} u \right] t + f(u) \qquad (2.88)$$

where $f(u)$ is the constant of integration. Note that at $t = 0$, $x = f(u)$.

Equations (2.84), (2.85), and (2.88) yield two solutions corresponding to waves traveling to the left ($-$) and right ($+$). From (2.83) and (2.86), it is evident that the characteristics are straight lines whose slopes depend on the value of u at the x intercept at $t = 0$. For each characteristic, the value of u is a constant.

The formation of a shock wave depends on the initial conditions. Consider a right traveling wave. From (2.87), the slope of a characteristic is

$$\left. \frac{\partial x}{\partial t} \right|_u = a_o + \frac{(\gamma + 1)}{2} u \qquad (2.89)$$

Thus, each point on the initial wave travels to the right with a velocity that

† This can always be assured by means of a Galilean transformation.

is a monotonically increasing function of u. Thus, if u decreases with increasing x over a portion of the initial wave, there will occur a crossing of the characteristics at some time t_s, implying a multiple-valued solution for $t > t_s$. Since this is physically unfeasible, a shock wave must occur at t_s which satisfies the shock conditions in Table 2.1. This is shown† schematically in Fig. 2.9 for an initial periodic profile for u.

At the instant t_s of the formation of the shock wave, the slope $\partial u / \partial x$ becomes infinite. At this point,

$$\frac{\partial x}{\partial u}\bigg|_t = 0 \quad \text{and} \quad \frac{\partial^2 x}{\partial u^2}\bigg|_t = 0 \tag{2.90}$$

From (2.88),

$$t_s = -\frac{2}{(\gamma + 1)}\frac{df}{du} \quad \text{and} \quad \frac{d^2 f}{du^2} = 0 \tag{2.91}$$

which defines the value of u at which (2.91) holds and the time t_s. The location of the shock formation is defined by (2.88). It is evident therefore that the formation of a shock(s) requires one or more inflection points u_s in the initial velocity distribution where $d^2 f / du^2 = 0$ and a negative slope df/du at $u = u_s$.

2.8 Shock Formation from Sinusoidal Disturbance

The following specific example will be employed in later chapters for evaluating various numerical algorithms. The initial condition for u is chosen to be

$$u(x,0) = \epsilon a_o \sin \kappa x \tag{2.92}$$

with $\kappa = 2\pi/\lambda$, where λ is the wavelength and ϵ is a dimensionless coefficient. The initial condition for the speed of sound a corresponding to a wave traveling to the right is

$$a = a_o + \frac{(\gamma - 1)}{2}u \tag{2.93}$$

The initial conditions for the remaining flow variables are obtained from (1.24) and (1.25). Now

$$f(u) = \kappa^{-1}\sin^{-1}\left(\frac{u}{\epsilon a_o}\right) \tag{2.94}$$

† Only one period is shown for clarity.

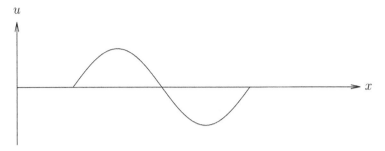

(a) Initial condition $(t = 0)$

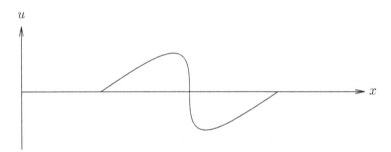

(b) At time of shock formation $(t = t_s)$

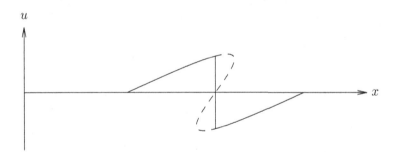

(c) After shock formation $(t > t_s)$

Fig. 2.9. Formation of shock wave

Taking into account the sign,

$$\frac{df}{du} = \begin{cases} -\kappa^{-1}\left[(\epsilon a_o)^2 - u^2\right]^{-1/2} & \text{for} \quad \frac{\pi}{2} \leq \kappa x \leq \frac{3\pi}{2} \\ \kappa^{-1}\left[(\epsilon a_o)^2 - u^2\right]^{-1/2} & \text{for} \quad -\frac{\pi}{2} \leq \kappa x \leq \frac{\pi}{2} \end{cases} \tag{2.95}$$

From (2.95),

$$\frac{d^2 f}{du^2} = 0 \quad \text{for} \quad u = 0 \tag{2.96}$$

and from (2.91),

$$t_s = \frac{2}{(\gamma + 1)\,\epsilon\kappa a_o} \tag{2.97}$$

From (2.88), the shock forms a distance $a_o t_s$ to the right of the point where $u = 0$ in the initial condition. Since the initial condition is periodic, a countably infinite number of shocks initially appear at t_s at locations $a_o t_s + n\lambda$ for $n = 0, \pm1, \ldots$.

2.9 General Riemann Problem

The General Riemann Problem,[†] illustrated in Fig. 2.10, is a useful paradigm for the development and testing of numerical algorithms for the one-dimensional Euler equations. At time $t = 0$, two different states of the flow exist and are separated by a contact surface at $x = 0$. For $x < 0$, the velocity u_1, static pressure p_1, and static temperature T_1 are given. For $x > 0$, the velocity u_4, static pressure p_4, and static temperature T_4 are likewise specified. Depending on the left and right states, there are four[‡] possible solutions for $t > 0$, which are illustrated in Figs. 2.12 to 2.15, namely, 1) two shock waves, 2) one shock wave and one expansion wave, 3) one expansion wave and one shock wave, and 4) two expansion waves. The solution of the General Riemann Problem can be constructed from the individual solutions for a propagating normal shock (Section 2.3) and an expansion wave (Section 2.5). In each case, a contact surface separates the fluid that was initially on either side of the interface at $t = 0$.

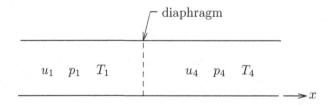

Fig. 2.10. Initial condition for the General Riemann Problem

The specific solution is determined by the contact surface pressure p^*,

---

[†] We assume identical values of the ratio of specific heats γ. The solution may be easily extended to allow for different γ on either side of the contact surface.

[‡] A fifth solution, comprised of left- and right-moving expansion waves and two contact surfaces with a vacuum in between (Gottlieb and Groth, 1988) is physically implausible.

which is defined by the transcendental equation§

$$a_1 f(p^*, p_1) + a_4 f(p^*, p_4) - u_1 + u_4 = 0 \qquad (2.98)$$

where

$$f(p^*, p) = \begin{cases} \frac{1}{\gamma} \left(\frac{p^*}{p} - 1 \right) \left[\frac{\gamma+1}{2\gamma} \frac{p^*}{p} + \frac{\gamma-1}{2\gamma} \right]^{-1/2} & \text{for } p^* \geq p \\ \frac{2}{(\gamma-1)} \left[\left(\frac{p^*}{p} \right)^{(\gamma-1)/2\gamma} - 1 \right] & \text{for } p^* \leq p \end{cases} \qquad (2.99)$$

and $a = \sqrt{\gamma R T}$ is the speed of sound. The behavior of $f(p^*, p)$ for $\gamma = 1.4$ is shown in Fig. 2.11.

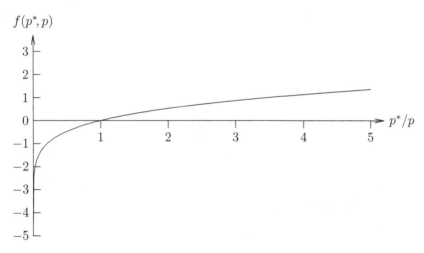

Fig. 2.11. Riemann function $f(p^*, p)$ for $\gamma = 1.4$

The four solutions are defined below and shown in Figs. 2.12 to 2.15. For each solution, there are four regions, each of which is separated by a simple wave (*e.g.*, shock wave, contact surface, or expansion wave). Although the simple waves are shown in Figs. 2.12 to 2.15 with a specific direction (*e.g.*, the left shock in Fig. 2.12 moves to the left), the waves may travel in either direction†.

§ The solution can be obtained using Newton's method. The initial guess may be taken to be (Gottlieb and Groth, 1988)

$$p^* = \frac{(\rho_1 a_1 p_4 + \rho_4 a_4 p_1 + \rho_1 \rho_4 a_1 a_4 (u_1 - u_4))}{(\rho_1 a_1 + \rho_4 a_4)}$$

Note that if $p^* < p_1$ and $p^* < p_4$ (Case 4), the equation may be solved directly:

$$p^* = \left[\left[(a_1 + a_4) + \frac{(\gamma-1)}{2} (u_1 - u_4) \right] \left[a_1 p_1^{-(\gamma-1)/2\gamma} + a_4 p_4^{-(\gamma-1)/2\gamma} \right]^{-1} \right]^{2\gamma/(\gamma-1)}$$

† Of course, the simple waves cannot cross each other.

2.9.1 Case 1. Two Shock Waves: $p_1 < p^*$ and $p_4 < p^*$

The velocities of the left and right shock waves (c_{s_l}) and (c_{s_r}) and contact surface (c_c) are[2]

$$c_{s_l} = u_1 - a_1 \sqrt{\frac{(\gamma + 1)}{2\gamma} \left(\frac{p^*}{p_1} - 1 \right) + 1} \qquad (2.100)$$

$$c_{s_r} = u_4 + a_4 \sqrt{\frac{(\gamma + 1)}{2\gamma} \left(\frac{p^*}{p_4} - 1 \right) + 1} \qquad (2.101)$$

$$c_c = u_1 - \frac{a_1}{\gamma} \left(\frac{p^*}{p_1} - 1 \right) \left[\frac{(\gamma + 1)}{2\gamma} \frac{p^*}{p_1} + \frac{(\gamma - 1)}{2\gamma} \right]^{-1/2} \qquad (2.102)$$

The flow variables in Region 2 $(c_{s_l} < x/t < c_c)$ are

$$u_2 = c_c \qquad (2.103)$$

$$p_2 = p^* \qquad (2.104)$$

$$\rho_2 = \rho_1 \left[(\gamma - 1) + (\gamma + 1)\frac{p^*}{p_1} \right] \left[(\gamma + 1) + (\gamma - 1)\frac{p^*}{p_1} \right]^{-1} \qquad (2.105)$$

The static temperature T_2 can be found from the ideal gas equation. The flow variables in Region 3 $(c_c < x/t < c_{s_r})$ are

$$u_3 = c_c \qquad (2.106)$$

$$p_3 = p^* \qquad (2.107)$$

$$\rho_3 = \rho_4 \left[(\gamma - 1) + (\gamma + 1)\frac{p^*}{p_4} \right] \left[(\gamma + 1) + (\gamma - 1)\frac{p^*}{p_4} \right]^{-1} \qquad (2.108)$$

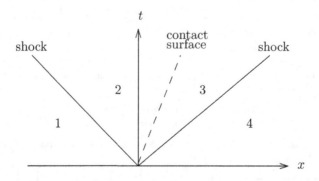

Fig. 2.12. General Riemann Problem: Case 1 (two shock waves)

2.9.2 Case 2. Shock and Expansion: $p_1 < p^*$ and $p_4 > p^*$

The velocities of the left shock (c_s), the left and right boundaries of the right expansion fan (c_l and c_r), and the contact surface (c_c) are[3]

$$c_s \;=\; u_1 - a_1 \sqrt{\frac{(\gamma+1)}{2\gamma}\left(\frac{p^*}{p_1} - 1\right) + 1} \tag{2.109}$$

$$c_l \;=\; u_4 + a_4 \left[\frac{(\gamma+1)}{(\gamma-1)}\left(\frac{p^*}{p_4}\right)^{(\gamma-1)/2\gamma} - \frac{2}{\gamma-1}\right] \tag{2.110}$$

$$c_r \;=\; u_4 + a_4 \tag{2.111}$$

$$c_c \;=\; u_1 - \frac{a_1}{\gamma}\left(\frac{p^*}{p_1} - 1\right)\left[\frac{(\gamma+1)}{2\gamma}\frac{p^*}{p_1} + \frac{(\gamma-1)}{2\gamma}\right]^{-1/2} \tag{2.112}$$

The flow variables in Region 2 are

$$u_2 \;=\; c_c \tag{2.113}$$

$$p_2 \;=\; p^* \tag{2.114}$$

$$\rho_2 \;=\; \rho_1 \left[(\gamma-1) + (\gamma+1)\frac{p^*}{p_1}\right]\left[(\gamma+1) + (\gamma-1)\frac{p^*}{p_1}\right]^{-1} \tag{2.115}$$

and the flow variables in Region 3 are

$$u_3 \;=\; c_c \tag{2.116}$$

$$p_3 \;=\; p^* \tag{2.117}$$

$$\rho_3 \;=\; \rho_4 \left(\frac{p^*}{p_4}\right)^{1/\gamma} \tag{2.118}$$

Within the expansion fan $c_l \le x/t \le c_r$,

$$u \;=\; \frac{2}{\gamma+1}\left[\frac{x}{t} + \frac{(\gamma-1)}{2}u_4 - a_4\right] \tag{2.119}$$

$$p \;=\; p_4 \left[\frac{(\gamma-1)}{(\gamma+1)}\frac{1}{a_4}\left(\frac{x}{t} - u_4\right) + \frac{2}{(\gamma+1)}\right]^{2\gamma/(\gamma-1)} \tag{2.120}$$

$$\rho \;=\; \rho_4 \left(\frac{p}{p_4}\right)^{1/\gamma} \tag{2.121}$$

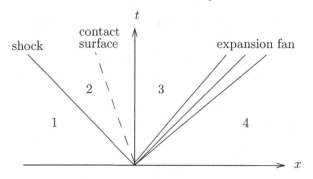

Fig. 2.13. General Riemann Problem: Case 2 (shock and expansion)

2.9.3 Case 3. Expansion and Shock: $p_1 > p^*$ and $p_4 < p^*$

The velocities of the right shock (c_s), the left and right boundaries of the left expansion fan (c_l and c_r), and the contact surface (c_c) are[4]

$$c_s = u_4 + a_4 \sqrt{\frac{(\gamma + 1)}{2\gamma} \left(\frac{p^*}{p_4} - 1 \right) + 1} \qquad (2.122)$$

$$c_l = u_1 - a_1 \qquad (2.123)$$

$$c_r = u_1 + a_1 \left[\frac{2}{\gamma - 1} - \frac{(\gamma + 1)}{(\gamma - 1)} \left(\frac{p^*}{p_1} \right)^{(\gamma - 1)/2\gamma} \right] \qquad (2.124)$$

$$c_c = u_4 + \frac{a_4}{\gamma} \left(\frac{p^*}{p_4} - 1 \right) \left[\frac{(\gamma + 1)}{2\gamma} \frac{p^*}{p_4} + \frac{(\gamma - 1)}{2\gamma} \right]^{-1/2} \qquad (2.125)$$

The flow variables in Region 2 are

$$u_2 = c_c \qquad (2.126)$$

$$p_2 = p^* \qquad (2.127)$$

$$\rho_2 = \rho_1 \left(\frac{p^*}{p_1} \right)^{1/\gamma} \qquad (2.128)$$

and the flow variables in Region 3 are

$$u_3 = c_c \qquad (2.129)$$

$$p_3 = p^* \qquad (2.130)$$

$$\rho_3 = \rho_4 \left[(\gamma - 1) + (\gamma + 1) \frac{p^*}{p_4} \right] \left[(\gamma + 1) + (\gamma - 1) \frac{p^*}{p_4} \right]^{-1} \qquad (2.131)$$

Within the expansion fan $c_l \le x/t \le c_r$,

$$u = \frac{2}{\gamma + 1} \left[\frac{x}{t} + \frac{(\gamma - 1)}{2} u_1 + a_1 \right] \qquad (2.132)$$

$$p = p_1 \left[\frac{(\gamma - 1)}{(\gamma + 1)} \frac{1}{a_1} \left(u_1 - \frac{x}{t} \right) + \frac{2}{(\gamma + 1)} \right]^{2\gamma/(\gamma-1)} \qquad (2.133)$$

$$\rho = \rho_1 \left(\frac{p}{p_1} \right)^{1/\gamma} \qquad (2.134)$$

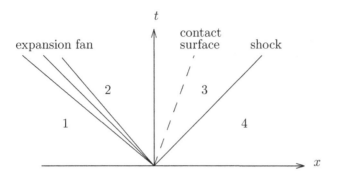

Fig. 2.14. General Riemann Problem: Case 3 (expansion and shock)

2.9.4 Case 4. Two Expansions: $p_1 > p^*$ and $p_4 > p^*$

The velocities of the left and right boundaries of the left expansion fan (c_{l_1} and c_{r_1}), the right expansion fan (c_{l_2} and c_{r_2}), and the contact surface (c_c) are[5]

$$c_{l_1} = u_1 - a_1 \qquad (2.135)$$

$$c_{r_1} = u_1 + a_1 \left[\frac{2}{\gamma - 1} - \frac{(\gamma + 1)}{(\gamma - 1)} \left(\frac{p^*}{p_1} \right)^{(\gamma-1)/2\gamma} \right] \qquad (2.136)$$

$$c_{l_2} = u_4 + a_4 \left[\frac{(\gamma + 1)}{(\gamma - 1)} \left(\frac{p^*}{p_4} \right)^{(\gamma-1)/2\gamma} - \frac{2}{\gamma - 1} \right] \qquad (2.137)$$

$$c_{r_2} = u_4 + a_4 \qquad (2.138)$$

$$c_c = u_1 + \frac{2a_1}{(\gamma - 1)} \left[1 - \left(\frac{p^*}{p_1} \right)^{(\gamma-1)/2\gamma} \right] \qquad (2.139)$$

The flow variables in Region 2 are

$$u_2 = c_c \qquad (2.140)$$

$$p_2 = p^* \qquad (2.141)$$

$$\rho_2 = \rho_1 \left(\frac{p^*}{p_1} \right)^{1/\gamma} \qquad (2.142)$$

and the flow variables in Region 3 are

$$u_3 = c_c \tag{2.143}$$

$$p_3 = p^* \tag{2.144}$$

$$\rho_3 = \rho_4 \left(\frac{p^*}{p_4}\right)^{1/\gamma} \tag{2.145}$$

Within the left expansion fan $c_{l_1} \leq x/t \leq c_{r_1}$,

$$u = \frac{2}{\gamma+1} \left[\frac{x}{t} + \frac{(\gamma-1)}{2} u_1 + a_1\right] \tag{2.146}$$

$$p = p_1 \left[\frac{(\gamma-1)}{(\gamma+1)} \frac{1}{a_1} \left(u_1 - \frac{x}{t}\right) + \frac{2}{(\gamma+1)}\right]^{2\gamma/(\gamma-1)} \tag{2.147}$$

$$\rho = \rho_1 \left(\frac{p}{p_1}\right)^{\frac{1}{\gamma}} \tag{2.148}$$

and within the right expansion fan $c_{l_2} \leq x/t \leq c_{r_2}$,

$$u = \frac{2}{\gamma+1} \left[\frac{x}{t} + \frac{(\gamma-1)}{2} u_4 - a_4\right] \tag{2.149}$$

$$p = p_4 \left[\frac{(\gamma-1)}{(\gamma+1)} \frac{1}{a_4} \left(\frac{x}{t} - u_4\right) + \frac{2}{(\gamma+1)}\right]^{2\gamma/(\gamma-1)} \tag{2.150}$$

$$\rho = \rho_4 \left(\frac{p}{p_4}\right)^{\frac{1}{\gamma}} \tag{2.151}$$

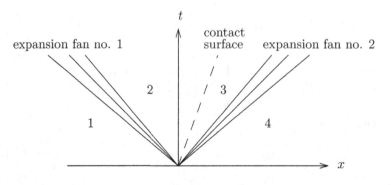

Fig. 2.15. General Riemann Problem: Case 4 (two expansions)

2.10 Riemann Shock Tube

The Riemann Shock Tube Problem is a special case of the General Riemann Problem wherein the initial velocities u_1 and u_4 are zero. We may assume, without loss of generality, that $p_4 > p_1$ and hence the configuration corresponds to Case 2. Equations (2.98) and (2.99) can then be rewritten as (Liepmann and Roshko, 1957)

$$\frac{p_4}{p_1} = \frac{p^*}{p_1} \left\{ 1 - \frac{(\gamma - 1)(a_1/a_4)(p^*/p_1 - 1)}{\sqrt{2\gamma}\sqrt{2\gamma + (\gamma + 1)(p^*/p_1 - 1)}} \right\}^{-2\gamma/(\gamma - 1)}$$

from which the shock pressure ratio $p_2/p_1 = p^*/p_1$ can be found by iteration. The shock speed is obtained from (2.109) as

$$c_s = -a_1 \sqrt{\frac{(\gamma + 1)}{2\gamma} \left(\frac{p^*}{p_1} - 1 \right) + 1}$$

The velocity of the left and right boundaries of the expansion fan are given by (2.110) and (2.111) as

$$c_l = a_4 \left[\frac{(\gamma + 1)}{(\gamma - 1)} \left(\frac{p^*}{p_4} \right)^{(\gamma - 1)/2\gamma} - \frac{2}{(\gamma - 1)} \right]$$
$$c_r = a_4$$

where $p^*/p_4 = (p^*/p_1)/(p_4/p_1)$. The velocity of the contact surface is obtained from (2.112) as

$$c_s = -\frac{a_1}{\gamma} \left(\frac{p^*}{p_1} - 1 \right) \left[\frac{(\gamma + 1)}{2\gamma} \frac{p^*}{p_1} + \frac{(\gamma - 1)}{2\gamma} \right]^{-1/2}$$

The flow variables in Regions 2 and 3 and the expansion fan are given by Equations (2.113) to (2.121).

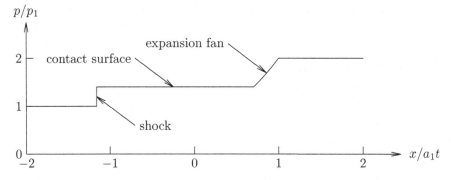

Fig. 2.16. Pressure for Riemann Shock Tube for $p_4/p_1 = 2$ and $T_4/T_1 = 1$

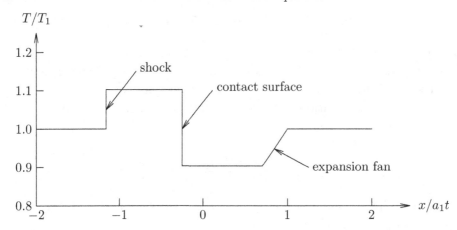

Fig. 2.17. Temperature for Riemann Shock Tube for $p_4/p_1 = 2$ and $T_4/T_1 = 1$

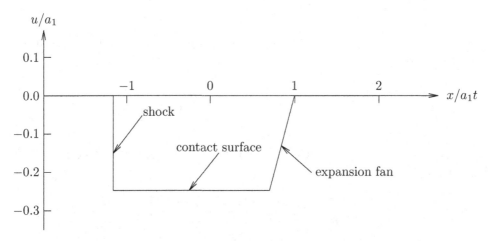

Fig. 2.18. Velocity for Riemann Shock Tube for $p_4/p_1 = 2$ and $T_4/T_1 = 1$

The static pressure and temperature, velocity, and entropy are shown in Figs. 2.16 to 2.19 for the initial conditions $p_4/p_1 = 2$ and $T_4/T_1 = 1$ with $\gamma = 1.4$. The abscissa is x/t normalized by a_1, whereby the solution at any time t can be obtained.† A shock moves to the left at velocity $c_s/a_1 = -1.159479$. The pressure, temperature, velocity, and entropy change discontinuously across the shock. At the contact surface, the pressure and velocity are continuous while the temperature (and hence the density) and entropy change discontinuously. The expansion fan moves to the right with left and right velocities $c_l/a_1 = 0.702978$ and $c_r/a_1 = 1.0$. Within the expansion fan, all flow variables change continuously.

† This is an example of a *conical* flow wherein there is no physical length scale.

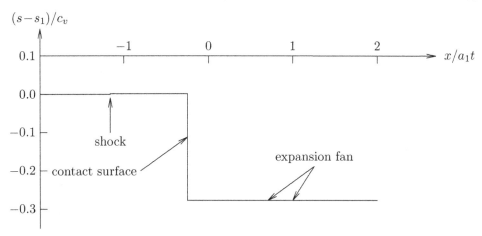

Fig. 2.19. Entropy for Riemann Shock Tube for $p_4/p_1 = 2$ and $T_4/T_1 = 1$

Exercises

2.1   Derive the Jacobian matrix (2.8).

SOLUTION

The Jacobian is obtained by differentiation of the flux vector $\mathcal{F} = (\mathcal{F}_1, \mathcal{F}_2, \mathcal{F}_3)^{\mathrm{T}}$ using (2.5). Denoting the element in the ith row and j^{th} column of $\mathcal{A}$ by $\mathcal{A}_{ij}$, we obtain

$$\mathcal{A}_{11} = \frac{\partial \mathcal{F}_1}{\partial \mathcal{Q}_1} = 0$$

$$\mathcal{A}_{12} = \frac{\partial \mathcal{F}_1}{\partial \mathcal{Q}_2} = 1$$

$$\mathcal{A}_{13} = \frac{\partial \mathcal{F}_1}{\partial \mathcal{Q}_3} = 0$$

$$\mathcal{A}_{21} = \frac{\partial \mathcal{F}_2}{\partial \mathcal{Q}_1} = \frac{(\gamma - 3)}{2} \left(\frac{\mathcal{Q}_2}{\mathcal{Q}_1} \right)^2$$

$$\mathcal{A}_{22} = \frac{\partial \mathcal{F}_2}{\partial \mathcal{Q}_2} = (3 - \gamma) \frac{\mathcal{Q}_2}{\mathcal{Q}_1}$$

$$\mathcal{A}_{23} = \frac{\partial \mathcal{F}_2}{\partial \mathcal{Q}_2} = (\gamma - 1)$$

$$\mathcal{A}_{31} = \frac{\partial \mathcal{F}_3}{\partial \mathcal{Q}_1} = -\gamma \frac{\mathcal{Q}_2 \mathcal{Q}_3}{\mathcal{Q}_1^2} + (\gamma - 1) \left(\frac{\mathcal{Q}_2}{\mathcal{Q}_1} \right)^3$$

$$\mathcal{A}_{32} = \frac{\partial \mathcal{F}_3}{\partial \mathcal{Q}_2} = \gamma \frac{\mathcal{Q}_3}{\mathcal{Q}_1} - \frac{3}{2} (\gamma - 1) \left(\frac{\mathcal{Q}_2}{\mathcal{Q}_1} \right)^2$$

$$\mathcal{A}_{33} = \frac{\partial \mathcal{F}_3}{\partial \mathcal{Q}_3} = \gamma \frac{\mathcal{Q}_2}{\mathcal{Q}_1}$$

2.2   Derive the eigenvalues of $\mathcal{A}$.

2.3   Derive the right eigenvectors of $\mathcal{A}$.

SOLUTION

A formal procedure for determining the eigenvectors can be defined using the property that the eigenvalues (2.12) are distinct (Franklin, 1968). Let $\mathcal{A}$ be an $n \times n$ matrix. Let

$\hat{\mathcal{A}}_q$ be the $(n-1) \times (n-1)$ matrix formed by crossing out row q and column q of $\mathcal{A}$, for $q = 1, \ldots, n$. At least one of the $\hat{\mathcal{A}}_q$ is nonsingular (Franklin, 1968). An eigenvector $e = (e_1, \ldots, e_n)^{\mathrm{T}}$ can be found by setting the qth element of e equal to 1 and solving the nonsingular system of equations

$$\sum_{\substack{j=1 \\ j \neq q}}^{n} (\mathcal{A}_{ij} - \lambda \delta_{ij}) \, e_j = -\mathcal{A}_{iq} \quad \text{for} \quad i = 1, \ldots, q-1, q+1, \ldots, n \qquad (\text{E2.1})$$

Consider the eigenvalue $\lambda = u$. Let $q = 1$. Then (E2.1) yields

$$\begin{aligned}
(\mathcal{A}_{22} - \lambda) \, e_2 + \mathcal{A}_{23} e_3 &= -\mathcal{A}_{21} \\
\mathcal{A}_{32} e_2 + (\mathcal{A}_{33} - \lambda) \, e_3 &= -\mathcal{A}_{31}
\end{aligned} \qquad (\text{E2.2})$$

The solution is $e_2 = u$ and $e_3 = \frac{1}{2}u^2$. Thus, the eigenvector corresponding to $\lambda_1 = u$ is $r_1 = (1, u, \frac{1}{2}u^2)^{\mathrm{T}}$, in agreement with (2.13). The remaining right eigenvectors are obtained in a similar manner.

2.4   Derive the left eigenvectors of $\mathcal{A}$

2.5   Prove Euler's Identity (2.24).

SOLUTION

Denoting $\mathcal{F} = \mathcal{A}\mathcal{Q}$ and using (2.8),

$$\begin{aligned}
\mathcal{F}_1 &= \mathcal{Q}_2 \\
\mathcal{F}_2 &= (\gamma-1)\mathcal{Q}_3 - \frac{(\gamma-3)}{2} \frac{\mathcal{Q}_2^2}{\mathcal{Q}_1} \\
\mathcal{F}_3 &= \frac{\gamma \mathcal{Q}_2 \mathcal{Q}_3}{\mathcal{Q}_1} - \frac{(\gamma-1)}{2} \frac{\mathcal{Q}_2^3}{\mathcal{Q}_1^2}
\end{aligned}$$

which is equal or equivalent to (2.5).

2.6   Is it true *in general* that the matrix T in a general similarity form (2.15) is the concatenation of the right eigenvectors of $\mathcal{A}$, and the matrix T^{-1} is the concatenation of the left eigenvectors of $\mathcal{A}$? Hint: Are the right and left eigenvectors unique?

2.7   For the case $u_1 - u_w > 0$ in Section 2.3, show that $s_2 - s_1 > 0$ implies $\sigma_l > 1$.

SOLUTION

From the definition of entropy,

$$s_2 - s_1 = c_v \log \left(\frac{p_2}{p_1} \right) - c_p \log \left(\frac{\rho_2}{\rho_1} \right)$$

Denote $\sigma = p_2/p_1$. From (2.42),

$$\frac{\rho_2}{\rho_1} = \frac{(\gamma - 1) + (\gamma + 1)\sigma}{(\gamma + 1) + (\gamma - 1)\sigma}$$

Thus

$$s_2 - s_1 = c_v \log \left\{ \left[\frac{(\gamma + 1) + (\gamma - 1)\sigma}{(\gamma - 1) + (\gamma + 1)\sigma} \right]^{\gamma} \sigma \right\}$$

Therefore, $s_2 - s_1 > 0$ implies that the argument of log must be greater than one. Since $\sigma > 0$ and $\gamma > 1$, we may take the $1/\gamma$ root of the argument and therefore

$$\left[\frac{(\gamma + 1) + (\gamma - 1)\sigma}{(\gamma - 1) + (\gamma + 1)\sigma} \right] \sigma^{1/\gamma} > 1$$

Let $\nu = \sigma^{1/\gamma}$. Then the above condition becomes

$$f(\nu) > 1$$

where

$$f(\nu) = \nu^{(\gamma+1)} - \frac{(\gamma+1)}{(\gamma-1)}\nu^\gamma + \frac{(\gamma+1)}{(\gamma-1)}\nu$$

Therefore

$$\frac{df}{d\nu} = (\gamma+1)\left[\nu^\gamma - \frac{\gamma}{(\gamma-1)}\nu^{(\gamma-1)} + \frac{1}{(\gamma-1)}\right]$$

$$\frac{d^2 f}{d\nu^2} = \gamma(\gamma+1)\nu^{(\gamma-2)}(\nu-1)$$

$$\frac{d^3 f}{d\nu^3} = \gamma(\gamma+1)\nu^{(\gamma-3)}[(\gamma-1)\nu - (\gamma-2)]$$

Thus, a Taylor's series expansion for $f(\nu)$ about $\nu = 1$ yields

$$f(\nu) = 1 + \frac{\gamma(\gamma+1)}{6}(\nu-1)^3 + \mathcal{O}((\nu-1)^4)$$

The Taylor series indicates that there exists a sufficiently small neighborhood of $\nu = 1$, where $f > 1$ if $\nu > 1$ and $f < 1$ if $\nu < 1$. For $\nu > 1$, $d^2 f/d\nu^2 > 0$. Thus, it is not possible for f to become less than one for $\nu > 1$, since this would require a change in sign of $d^2 f/d\nu^2$. By a similar argument, $f < 1$ for $\nu < 1$. Thus, $f > 1$ requires $\nu > 1$ and hence $s_2 - s_1 > 0$ requires $\sigma_l > 1$.

2.8 Redraw Figs. 2.6 and 2.7 for the other three possible cases: 1) $u > a$, 2) $-a < u < 0$, and 3) $u < -a$.

2.9 Determine the solution to the General Riemann Problem corresponding to a vacuum and two expansion fans.

SOLUTION

The configuration is illustrated in the figure below. The solution in the left expansion fan is

$$u = \frac{2}{(\gamma+1)}\left[\frac{x}{t} + \frac{(\gamma-1)}{2}u_1 + a_1\right]$$

$$a = \frac{(\gamma-1)}{(\gamma+1)}\left[u_1 - \frac{x}{t}\right] + \frac{2}{(\gamma+1)}a_1$$

$$\frac{p}{p_1} = \left(\frac{a}{a_1}\right)^{2\gamma/(\gamma-1)}$$

which is valid in the region $u_1 - a_1 \leq x/t \leq u_2 - a_2$. By assumption, there is a vacuum adjacent to the right boundary of the left expansion fan, and therefore $a_2 = 0$. Thus, from above,

$$u_2 = u_1 + \frac{2}{(\gamma-1)}a_1$$

and hence the left expansion fan is the region

$$u_1 - a_1 \leq \frac{x}{t} \leq u_1 + \frac{2}{(\gamma-1)}a_1$$

Similarly, the solution in the right expansion fan is

$$u = \frac{2}{(\gamma+1)}\left[\frac{x}{t} + \frac{(\gamma-1)}{2}u_4 - a_4\right]$$

$$a = \frac{(\gamma-1)}{(\gamma+1)}\left[\frac{x}{t} - u_4\right] + \frac{2}{(\gamma+1)}a_4$$

$$\frac{p}{p_4} = \left(\frac{a}{a_4}\right)^{2\gamma/(\gamma-1)}$$

which is valid in the region $u_3 + a_3 \le x/t \le u_4 + a_4$. By assumption, there is a vacuum adjacent to the right boundary of the left expansion fan, and therefore $a_3 = 0$. Thus, from above,

$$u_3 = u_4 - \frac{2}{(\gamma-1)}a_4$$

and hence the left expansion fan is the region

$$u_4 - \frac{2}{(\gamma-1)}a_4 \le \frac{x}{t} \le u_4 + a_4$$

The contact surfaces coincide with the inner boundaries of the expansion fans. The solution exists provided that the contact surfaces are separated,

$$u_4 - u_1 > \frac{2}{(\gamma-1)}(a_1 + a_4)$$

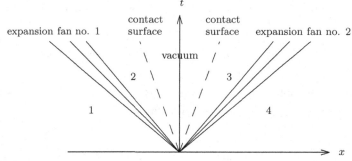

2.10   Solve the General Riemann Problem using the linearized Euler equations

$$\frac{\partial \mathcal{Q}}{\partial t} + \bar{\mathcal{A}}\frac{\partial \mathcal{Q}}{\partial x} = 0$$

where

$$\bar{\mathcal{A}} = \left\{ \begin{array}{ccc} 0 & 1 & 0 \\ (\gamma-1)\bar{u}^2/2 & (3-\gamma)\bar{u} & (\gamma-1) \\ -\bar{H}\bar{u}+(\gamma-1)\bar{u}^3/2 & \bar{H}-(\gamma-1)\bar{u}^2 & \gamma\bar{u} \end{array} \right\}$$

where

$$\bar{u} = \frac{u_1 + u_4}{2}$$

$$\bar{H} = \frac{H_1 + H_4}{2}$$

3

Accuracy, Consistency, Convergence, and Stability

It is perfectly obvious that a revolution of some sort became due in the field of numerical methods with the advent of modern computing machines.

Robert Richtmeyer (1967)

3.1 Introduction

The chapter presents the concepts of accuracy, consistency, convergence, and stability in the framework of the one-dimensional unsteady Euler equations. A simple discretization is derived for the purposes of illustration. Practical discretization methods are presented in subsequent chapters.

3.2 The Problem

The fundamental problem is to define an accurate algorithm for integrating the one-dimensional Euler equations in control volume form:

$$\frac{d}{dt} \int_V \mathcal{Q} \, dx dy + \int_{\partial V} \mathcal{F} dy = 0 \tag{3.1}$$

where $\mathcal{Q}$ is the vector of dependent variables,

$$\mathcal{Q} = \left\{ \begin{array}{c} \mathcal{Q}_1 \\ \mathcal{Q}_2 \\ \mathcal{Q}_3 \end{array} \right\} = \left\{ \begin{array}{c} \rho \\ \rho u \\ \rho e \end{array} \right\} \tag{3.2}$$

45

and $\mathcal{F}$ is the vector of fluxes

$$\mathcal{F} = \left\{ \begin{array}{c} \mathcal{F}_1 \\ \mathcal{F}_2 \\ \mathcal{F}_3 \end{array} \right\} = \left\{ \begin{array}{c} \rho u \\ \rho u u + p \\ \rho e u + p u \end{array} \right\} \tag{3.3}$$

3.3 Discretization

The conservation equations (3.1) are applied to a discrete set of control volumes V_i. The solution requires specification of a set of control volumes and algorithms for the temporal and spatial quadratures.

Assume a uniform discretization of the x-axis into M cells of length Δx with centroids x_i, $i = 1, \ldots, M$ as illustrated in Fig. 3.1. The cell faces, located midway between the adjacent centroids, are denoted by $x_{i+\frac{1}{2}}$ for $i = 0, \ldots, M$.

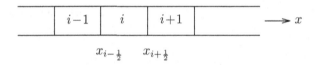

Fig. 3.1. Grid of control volumes V_{i-1}, V_i, V_{i+1}

Assume a discretization of time into discrete levels t^n, $n = 1, \ldots$, where

$$t^{n+1} = t^n + \Delta t^n \tag{3.4}$$

For volume i, denote the volume-averaged vector of dependent variables by

$$Q_i(t) = \frac{1}{V_i} \int_{V_i} \mathcal{Q} \, dxdy \tag{3.5}$$

where $V_i = \Delta x \Delta y$ and Δy is the (constant) height of each cell. The spatial (flux) quadrature involves faces $i + \frac{1}{2}$ and $i - \frac{1}{2}$. Denote

$$F_{i+\frac{1}{2}} = \frac{1}{A_{i+\frac{1}{2}}} \int_{x_{i+\frac{1}{2}}} \mathcal{F} dy \tag{3.6}$$

where $A_{i+\frac{1}{2}} = \Delta y$ is the surface area of the face at $x_{i+\frac{1}{2}}$. The flux vector $\mathcal{F}$ depends on $\mathcal{Q}$. In the discretization, the dependence on $\mathcal{Q}$ can be replaced by an assumed dependence on $Q_{i+\frac{1}{2}}$, which is some function f of a set of the volume-averaged variables Q_i in the neighborhood of $x_{i+\frac{1}{2}}$, namely,

$$Q_{i+\frac{1}{2}} = f\left(Q_{i-m}, Q_{i-m+1}, \ldots, Q_i, \ldots, Q_{i+n-1}, Q_{i+n}\right) \tag{3.7}$$

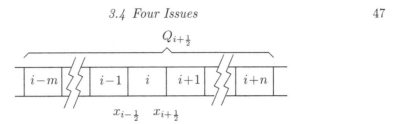

Fig. 3.2. Domain of dependence for $Q_{i+\frac{1}{2}}$

as illustrated in Fig. 3.2.

Then the Euler equations (3.1) become

$$\frac{dQ_i V_i}{dt} + \left(F_{i+\frac{1}{2}}\Delta y - F_{i-\frac{1}{2}}\Delta y\right) = 0 \qquad (3.8)$$

Since V_i is assumed to be independent of time,

$$\frac{dQ_i}{dt} + \frac{\left(F_{i+\frac{1}{2}} - F_{i-\frac{1}{2}}\right)}{\Delta x} = 0 \qquad (3.9)$$

This is the *semi-discrete* method by which the Euler partial differential equations (2.1) are transformed into a system of ordinary differential equations. This approach is also known as the *Method of Lines* (Hirsch, 1988; Fletcher, 1988; Holt 1984).

Given the solution Q_i for $i = 1, \ldots, M$ at time t^n, the solution at time t^{n+1} can be obtained by integration:

$$Q_i^{n+1} = Q_i^n - \frac{1}{\Delta x} \int_{t^n}^{t^{n+1}} \left(F_{i+\frac{1}{2}} - F_{i-\frac{1}{2}}\right) dt \qquad (3.10)$$

The problem therefore reduces to defining the *temporal* and *spatial* quadrature algorithms for evaluating the second term on the right side of (3.10).

3.4 Four Issues

In developing algorithms for the solution of (3.10), there are four issues which must be addressed (Richtmeyer and Morton, 1967; Anderson *et al.*, 1984; Hoffman, 1989; Morton and Mayers, 1994; Toro, 1997). The first is *accuracy*, which is the quantitative measure of the agreement between the numerical simulation and the exact solution on a given grid. The second is *consistency*, which is the fidelity of the numerical simulation in representing the actual solution of the governing partial differential equations. The third is *stability*, which represents the absence of temporally unbounded oscillations of a nonphysical nature in the simulation. The fourth is *convergence*, which is the property that the numerical solution converges to a solution when the spatial and temporal grid spacings are reduced to arbitrarily small values.

These four issues will be made more precise in subsequent sections. First, we present a class of discrete approximations to (3.10) for the purposes of examining accuracy, consistency, stability, and convergence. Second, we consider the accuracy of this specific class of discrete approximations. Third, we determine the necessary conditions for the consistency of this class of discrete approximations. Fourth, we examine the effect of the spatial flux quadrature algorithm (3.6) on the stability of (3.9) viewed as an ordinary differential equation, *i.e.*, without discretization of the time derivative. We show that the spatial flux quadrature algorithm must properly represent the physics of the flow to prevent the occurrence of unphysical oscillations that grow unbounded in time. Additionally, specific problems associated with the formation of shock waves are identified. We assess the numerical stability of the class of discrete approximations introduced earlier. Finally, we consider the issue of convergence.

3.5 A Class of Discrete Approximations

We consider a class of discrete approximations to (3.10):

$$Q_i^{n+1} = Q_i^n - \frac{1}{\Delta x} \int_{t_n}^{t^{n+1}} \left(F_{i+\frac{1}{2}} - F_{i-\frac{1}{2}} \right) dt$$

where, according to (3.6),

$$F_{i+\frac{1}{2}} = \frac{1}{\Delta y} \int_{x_{i+\frac{1}{2}}} \mathcal{F} dy$$

and thus

$$F_{i+\frac{1}{2}} = F(\mathcal{Q}_{i+\frac{1}{2}}) = \mathcal{F}(\mathcal{Q}_{i+\frac{1}{2}})$$

The class is limited in scope for the purposes of discussing the issues of accuracy, consistency, stability, and convergence, and extensions are presented in later chapters.

The approximation proceeds in two steps. First, we consider a two-level approximation to the time integral,

$$\int_{t_n}^{t^{n+1}} F_{i+\frac{1}{2}} \, dt \Rightarrow \left[\theta F(\mathcal{Q}_{i+\frac{1}{2}}^{n+1}) + (1-\theta)F(\mathcal{Q}_{i+\frac{1}{2}}^n) \right] \Delta t \qquad (3.11)$$

where θ is a constant and $\Rightarrow$ indicates that the integral is replaced by the discrete expression on the right. This is the *generalized trapezoidal method.* Second, we replace $\mathcal{Q}_{i+\frac{1}{2}}$ in (3.11) by

$$\mathcal{Q}_{i+\frac{1}{2}} \Rightarrow Q_{i+\frac{1}{2}} \qquad (3.12)$$

where $Q_{i+\frac{1}{2}}$ is defined according to (3.7) as

$$Q_{i+\frac{1}{2}} = f\left(Q_{i-m}, Q_{i-m+1}, \ldots, Q_i, \ldots, Q_{i+n-1}, Q_{i+n}\right)$$

and the specific form of the function f is yet to be determined (see Section 3.8).

The discrete approximation to (3.10) is therefore

$$\begin{aligned}
Q_i^{n+1} = Q_i^n \quad &- \quad \frac{\Delta t}{\Delta x} \left\{ \theta \left[F(Q_{i+\frac{1}{2}}^{n+1}) - F(Q_{i-\frac{1}{2}}^{n+1}) \right] \right\} \\
&- \frac{\Delta t}{\Delta x} \left\{ (1-\theta) \left[F(Q_{i+\frac{1}{2}}^n) - F(Q_{i-\frac{1}{2}}^n) \right] \right\} \qquad (3.13)
\end{aligned}$$

3.6 Accuracy

In this section, we determine the accuracy of the specific class of discrete approximations (3.13). We define the error $\mathcal{E}$ in approximating (3.10) by (3.13):

$$Q_i^{n+1} \Big|_{\text{discrete}} = Q_i^{n+1} \Big|_{\text{exact}} + \mathcal{E}(\Delta t, \Delta x) \qquad (3.14)$$

where $Q_i^{n+1} \Big|_{\text{exact}}$ is defined by (3.10) and $\mathcal{E}$ depends on Δx and Δt. We anticipate† that $\mathcal{E}$ can be expanded in a Taylor series,

$$\begin{aligned}
\mathcal{E}(\Delta x, \Delta t) \quad = \quad & a\Delta x + b\Delta t + c\Delta x^2 + d\Delta x \Delta t + e\Delta t^2 + \\
& \mathcal{O}(\Delta x^i \Delta t^{3-i}) \Big|_{i=0,\ldots,3} + \cdots \qquad (3.15)
\end{aligned}$$

where the coefficients $a, \ldots, e$ are functions of x and t. The notation for the remainder $\mathcal{O}(\Delta x^i \Delta t^{3-i})$ indicates that the term is proportional‡ to

† See Section 3.7.
‡ More precisely,

$$f(x) = \mathcal{O}(g(x)) \quad \text{as} \quad x \to x_o \quad \text{if} \quad \lim_{x \to x_o} \frac{f(x)}{g(x)} = \alpha$$

$\Delta x^i \Delta t^{3-i}$. The leading terms of $\mathcal{E}$ are $\mathcal{O}(\Delta x)$ and $\mathcal{O}(\Delta t)$ since $\mathcal{E}$ must vanish as Δx and Δt approach zero for any reasonable discretization.

The first step is to replace Q in (3.13) by an expression involving $\mathcal{Q}$. We can assume that

$$Q_{i+\frac{1}{2}} = \left[\mathcal{Q}(x) + \Delta x \mathcal{R}_1(x) + \Delta x^2 \mathcal{R}_2(x) + \mathcal{O}(\Delta x^3) \right]_{x_{i+\frac{1}{2}}} \tag{3.16}$$

where $\mathcal{R}_1$ and $\mathcal{R}_2$ depend on the function f in (3.7). Using a Taylor expansion†

$$
\begin{aligned}
F(Q_{i+\frac{1}{2}}) &= \left. F(\mathcal{Q}(x) + \Delta x \mathcal{R}_1(x) + \Delta x^2 \mathcal{R}_2(x) + \mathcal{O}(\Delta x^3)) \right|_{x_{i+\frac{1}{2}}} \\
&= F(\mathcal{Q}(x)) + \frac{\partial F}{\partial Q} \left[\Delta x \mathcal{R}_1(x) + \Delta x^2 \mathcal{R}_2(x) + \mathcal{O}(\Delta x^3) \right] \\
&\quad + \mathcal{O}(\Delta x^2)
\end{aligned}
$$

where the right side is evaluated at $x_{i+\frac{1}{2}}$.

The second step is to expand the flux difference in (3.13):

$$
\begin{aligned}
F(Q^n_{i+\frac{1}{2}}) - F(Q^n_{i-\frac{1}{2}}) &= F(\mathcal{Q}^n_{i+\frac{1}{2}}) - F(\mathcal{Q}^n_{i-\frac{1}{2}}) \\
&\quad + \Delta x^2 \frac{\partial}{\partial x} \left(\frac{\partial F}{\partial Q} \mathcal{R}_1 \right) + \mathcal{O}(\Delta x^3) \tag{3.17}
\end{aligned}
$$

The third step is to substitute (3.17) into the discrete approximation (3.13), which yields

$$
\begin{aligned}
Q^{n+1}_i = Q^n_i \;&-\; \frac{\Delta t}{\Delta x} \left\{ \theta \left[F(\mathcal{Q}^{n+1}_{i+\frac{1}{2}}) - F(\mathcal{Q}^{n+1}_{i-\frac{1}{2}}) \right] \right\} \\
&-\; \frac{\Delta t}{\Delta x} \left\{ (1 - \theta) \left[F(\mathcal{Q}^n_{i+\frac{1}{2}}) - F(\mathcal{Q}^n_{i-\frac{1}{2}}) \right] \right\} \\
&-\; \frac{\Delta t}{\Delta x} \left\{ \Delta x^2 \left[\theta \frac{\partial}{\partial x} \left(\frac{\partial F}{\partial Q} \mathcal{R}_1 \right)^{n+1} + (1 - \theta) \frac{\partial}{\partial x} \left(\frac{\partial F}{\partial Q} \mathcal{R}_1 \right)^n \right] \right\} \\
&+\; \mathcal{O}(\Delta x^2 \Delta t) \tag{3.18}
\end{aligned}
$$

The fourth step is to incorporate the error in the time discretization (3.11). It may be shown (Mathews and Fink, 1998) that

$$\left[\theta F(\mathcal{Q}^{n+1}_{i+\frac{1}{2}}) + (1 - \theta) F(\mathcal{Q}^n_{i+\frac{1}{2}}) \right] \Delta t =$$

where α is nonzero and finite. An additional notation is

$$f(x) = o(g(x)) \quad \text{as} \quad x \to x_o \quad \text{if} \quad \lim_{x \to x_o} \frac{f(x)}{g(x)} = 0$$

† Note that $\mathcal{R}_1$ and $\mathcal{R}_2$ are vectors and $\partial F / \partial Q$ is a matrix.

$$\int_{t_n}^{t^{n+1}} F_{i+\frac{1}{2}}\, dt + (\theta - \tfrac{1}{2})\Delta t^2 T_1(x_{i+\frac{1}{2}}) + \mathcal{O}(\Delta t^3) \qquad (3.19)$$

where T_1 is expressed in terms of F and θ. Substituting into (3.18),

$$
\begin{aligned}
Q_i^{n+1} = Q_i^n \ &- \ \frac{1}{\Delta x} \int_{t_n}^{t^{n+1}} \Big(F(\mathcal{Q}_{i+\frac{1}{2}}) - F(\mathcal{Q}_{i-\frac{1}{2}}) \Big)\, dt \\
&- \ (\theta - \tfrac{1}{2})\Delta t^2 \frac{\partial T_1}{\partial x} + \mathcal{O}(\Delta t^2 \Delta x) + \mathcal{O}(\Delta t^3) \\
&- \ \Delta x \Delta t \left[\theta \frac{\partial}{\partial x}\left(\frac{\partial F}{\partial Q} \mathcal{R}_1 \right)^{n+1} + (1-\theta)\frac{\partial}{\partial x}\left(\frac{\partial F}{\partial Q} \mathcal{R}_1 \right)^n \right] \\
&+ \ \mathcal{O}(\Delta x^2 \Delta t) \qquad\qquad\qquad\qquad\qquad\qquad (3.20)
\end{aligned}
$$

The discrete system (3.13) is therefore equivalent to

$$\underbrace{Q_i^{n+1} = Q_i^n - \frac{1}{\Delta x}\int_{t_n}^{t^{n+1}}\Big(F_{i+\frac{1}{2}} - F_{i-\frac{1}{2}} \Big)\, dt}_{Q_i^{n+1}|_{\text{exact}}} + \mathcal{E}(\Delta x, \Delta t) \qquad (3.21)$$

where the coefficients (3.15) in the error $\mathcal{E}(\Delta x, \Delta t)$ are

$$
\begin{aligned}
a &= 0 \\
b &= 0 \\
c &= 0 \\
d &= -\left[\theta \frac{\partial}{\partial x}\left(\frac{\partial F}{\partial Q}\mathcal{R}_1 \right)^{n+1} + (1-\theta)\frac{\partial}{\partial x}\left(\frac{\partial F}{\partial Q}\mathcal{R}_1 \right)^n \right] \\
e &= -(\theta - \tfrac{1}{2})\frac{\partial T_1}{\partial x} \qquad\qquad\qquad\qquad\qquad\qquad (3.22)
\end{aligned}
$$

The error is described in terms of its temporal and spatial components. If $\theta = \tfrac{1}{2}$, the algorithm is *temporally second-order accurate*, and if $\theta \neq \tfrac{1}{2}$, it is *temporally first-order accurate*.† If $\mathcal{R}_1 = 0$, then the algorithm is *spatially second-order accurate*, and if $\mathcal{R}_1 \neq 0$, then it is *spatially first-order accurate*.

3.7 Consistency

The discrete approximation (3.13) is *consistent* with the exact equation (3.10) if it yields the integral equation (3.10) in the limit of $\Delta t \to 0$ and

---

† Three typical values are $\theta = \tfrac{1}{2}$ (*Crank-Nicholson* or *trapezium*), $\theta = 0$ (*explicit Euler*), and $\theta = 1$ (*implicit Euler*).

the differential equation (2.1) in the limit of $\Delta x \to 0$ and $\Delta t \to 0$. This is determined by the nature of the error $\mathcal{E}(\Delta x, \Delta t)$.

From (3.14),

$$Q_i^{n+1}\Big|_{\text{discrete}} = Q_i^{n+1}\Big|_{\text{exact}} + \mathcal{E}(\Delta t, \Delta x)$$

where, from (3.15), we have

$$\mathcal{E}(\Delta x, \Delta t) = a\Delta x + b\Delta t + c\Delta x^2 + d\Delta x\Delta t + e\Delta t^2 +$$
$$\mathcal{O}(\Delta x^i \Delta t^{3-i})\Big|_{i=0,\dots,3} + \dots$$

and from (3.10) we have

$$Q_i^{n+1} = Q_i^n - \frac{1}{\Delta x}\int_{t_n}^{t^{n+1}}\left(F_{i+\frac{1}{2}} - F_{i-\frac{1}{2}}\right) dt$$

Assume the values for Q_i^{n+1} are exact and that the discrete expression (3.13) is used to determine $Q_i^{n+1}|_{\text{discrete}}$. Then

$$\frac{Q_i^{n+1}\Big|_{\text{discrete}} - Q_i^n|_{\text{exact}}}{\Delta t} + \frac{1}{\Delta t\Delta x}\int_{t_n}^{t^{n+1}}\left(F_{i+\frac{1}{2}} - F_{i-\frac{1}{2}}\right) dt = \frac{\mathcal{E}}{\Delta t} \qquad (3.23)$$

Taking the limit $\Delta t \to 0$, this must approach (3.9) as

$$\frac{dQ_i}{dt} + \frac{\left(F_{i+\frac{1}{2}} - F_{i-\frac{1}{2}}\right)}{\Delta x} = 0$$

which implies

$$\lim_{\substack{\Delta t \to 0 \\ \Delta x \text{ fixed}}} \frac{\mathcal{E}}{\Delta t} = 0 \qquad (3.24)$$

Thus, the discrete approximation (3.13) is consistent with the integral equation (3.10) provided

$$a = 0$$
$$b = 0$$
$$c = 0$$
$$\mathcal{O}(\Delta x^i \Delta t^{3-i})\Big|_{i=0,\dots,3} \quad \text{is} \quad \mathcal{O}(\Delta x^i \Delta t^{3-i})\Big|_{i=0,\dots,1} \qquad (3.25)$$

and so forth.

The additional requirement is consistency with the differential equation (2.1). Taking the limit of (3.23) as $\Delta t \to 0$ and $\Delta x \to 0$, the condition is

$$\lim_{\substack{\Delta t \to 0 \\ \Delta x \to 0}} \frac{\mathcal{E}}{\Delta t} = 0 \qquad (3.26)$$

which is already satisfied by (3.25). Thus, the discrete approximation (3.13) is consistent[1] with (3.10).

3.8 Flux Quadrature and Stability

The choice of spatial discretization for (3.9) has a significant impact on the accuracy of the solution. Certain discretizations will yield exponentially growing solutions that are physically implausible.† Such discretications are denoted *unstable*. Similarly, discretizations that avoid such exponentially growing solutions are denoted *stable*.

In the following subsections, we present two different flux quadrature methods. These methods are chosen for their obvious simplicity‡ to illustrate a fundamental requirement, namely, *the spatial flux quadrature must respect the physical domain of dependence*. To this end, we show by example that, irrespective of the temporal quadrature method, the failure to meet this requirement leads to exponentially growing perturbations and an unsatisfactory solution. We also show that satisfying this requirement, although necessary, is not sufficient to guarantee an acceptable solution everywhere (Section 3.8.4).

3.8.1 A Simple Flux Quadrature

We begin with a simple and straightforward approach to illustrate the concept. We make two assumptions, namely,

$$\begin{aligned}
F_{i+\frac{1}{2}} &= \mathcal{F}(Q_{i+\frac{1}{2}}) \\
Q_{i+\frac{1}{2}} &= \tfrac{1}{2}\left(Q_i + Q_{i+1}\right)
\end{aligned} \tag{3.27}$$

The first assumption states that the flux quadrature is based on an interpolated value $Q_{i+\frac{1}{2}}$, *i.e.*, an approximate value for $\mathcal{Q}$ is obtained at $x_{i+\frac{1}{2}}$ from (3.7) which is used to define $F_{i+\frac{1}{2}}$ using (2.5). The second assumption defines the interpolation (3.7), *i.e.*, $Q_{i+\frac{1}{2}}$ is the simple spatial average.

We seek to determine the stability of (3.27) using the analysis developed by von Neumann (1950)[2] . For purposes of simplicity, we assume that the

---

† Here we distinguish purely numerical instabilities from physically unstable solutions of the linearized equations of motion as those that arise in hydrodynamic stability theory (see, for example, Chandrasekhara, 1981).

‡ Albeit they would not likely be employed in any practical computation.

flow is periodic in x over a length $L = (M-1)\Delta x$,

$$Q_1 = Q_M \qquad (3.28)$$

and assume M is odd with $M = 2N+1$. Consider $Q(x,t)$ to be a continuous vector function that interpolates Q_i:

$$Q(x_i, t) = Q_i(t) \qquad (3.29)$$

Then the Fourier series for $Q(x,t)$ is

$$Q(x,t) = \sum_{l=-N+1}^{l=N} \hat{Q}_k(t) e^{\iota k x} \qquad (3.30)$$

where $\iota = \sqrt{-1}$ and the wavenumber k depends on l according to

$$k = \frac{2\pi l}{L} \qquad (3.31)$$

The Fourier coefficients $\hat{Q}_k(t)$ are complex vectors whose subscript k indicates an ordering with respect to the summation index l, *i.e.*, $\hat{Q}_k(t)$ indicates dependence on l (through (3.31)) and on t. Given the values of Q_i at some time t^n, the Fourier coefficients $\hat{Q}_k$ at t^n are obtained from

$$\hat{Q}_k(t^n) = \frac{1}{2N} \sum_{i=1}^{i=2N} Q_i(t^n) e^{-\iota k x_i} \qquad (3.32)$$

Consider the discrete Euler equations

$$\frac{dQ_i}{dt} + \frac{\left(F_{i+\frac{1}{2}} - F_{i-\frac{1}{2}}\right)}{\Delta x} = 0 \qquad (3.33)$$

The flux vector $F_{i+\frac{1}{2}}$ is expanded in a Taylor series about Q_i,

$$F_{i+\frac{1}{2}} = F(Q_{i+\frac{1}{2}}) = F(Q_i) + \left.\frac{\partial F}{\partial Q}\right|_i \left(Q_{i+\frac{1}{2}} - Q_i\right) + \mathcal{O}\left((\Delta Q)^2\right) \qquad (3.34)$$

where we have substituted $\mathcal{F}$ for F using (3.27). The term $\partial F/\partial Q|_i$ is denoted A_i and is $\mathcal{A}(Q_i)$.

The quantity $\mathcal{O}((\Delta Q)^2)$ represents the terms that are quadratic in the difference $\Delta Q = Q_{i+\frac{1}{2}} - Q_i$. Thus, truncating (3.34) to linear terms,

$$F_{i+\frac{1}{2}} \approx F(Q_i) + A_i \left(Q_{i+\frac{1}{2}} - Q_i\right) \qquad (3.35)$$

where the $\approx$ sign will hereafter be replaced by an equals sign. Using (3.27),

$$F_{i+\frac{1}{2}} = F(Q_i) + \tfrac{1}{2}A_i \left(Q_{i+1} - Q_i\right) \qquad (3.36)$$

and similarly

$$F_{i-\frac{1}{2}} = F(Q_i) - \tfrac{1}{2} A_i \left(Q_i - Q_{i-1} \right) \tag{3.37}$$

Thus

$$F_{i+\frac{1}{2}} - F_{i-\frac{1}{2}} = \tfrac{1}{2} A_i \left(Q_{i+1} - Q_{i-1} \right) \tag{3.38}$$

and (3.33) becomes

$$\frac{dQ_i}{dt} + A_i \frac{(Q_{i+1} - Q_{i-1})}{2\Delta x} = 0 \tag{3.39}$$

Assume A is slowly varying in x and hence can be treated as a constant to a first approximation. Substituting for Q_i from the (interpolating) Fourier series (3.30),

$$\sum_{l=-N+1}^{l=N} \frac{d\hat{Q}_k}{dt} e^{\iota k x_i} + \frac{1}{2\Delta x} A \sum_{l=-N+1}^{l=N} \hat{Q}_k \left[e^{\iota k(x_i + \Delta x)} - e^{\iota k(x_i - \Delta x)} \right] \tag{3.40}$$

The coefficients of the individual Fourier terms $e^{\iota k x_i}$ must vanish, and therefore

$$\frac{d\hat{Q}_k}{dt} + G\hat{Q}_k = 0 \quad \text{for } l = -N+1, \dots, N \tag{3.41}$$

or equivalently

$$\frac{d\hat{Q}_k}{dt} + G\hat{Q}_k = 0 \quad \text{for } k = \frac{\pi}{\Delta x}\frac{(-N+1)}{N}, \dots, \frac{\pi}{\Delta x} \tag{3.42}$$

where

$$G = \alpha A \tag{3.43}$$

implying that each element of the matrix A is multiplied by the complex coefficient α defined as

$$\alpha = \frac{\iota \sin k\Delta x}{\Delta x} \tag{3.44}$$

Equation (3.41) can be solved exactly. Using (2.15),

$$A = T\Lambda T^{-1} \tag{3.45}$$

Multiplying (3.41) by T^{-1} and defining

$$\tilde{Q}_k = T^{-1}\hat{Q}_k \tag{3.46}$$

gives

$$\frac{d\tilde{Q}_k}{dt} + \alpha\Lambda\tilde{Q}_k = 0 \tag{3.47}$$

Defining

$$\tilde{Q}_k = \left\{ \begin{array}{c} \tilde{Q}_{k_1} \\ \tilde{Q}_{k_2} \\ \tilde{Q}_{k_3} \end{array} \right\} \tag{3.48}$$

gives

$$\tilde{Q}_{k_m}(t) = \tilde{Q}_{k_m}(0)e^{-\omega_m t} \tag{3.49}$$

for $m = 1, 2, 3$ and $l = -N+1, \ldots, N$, where ω_m are complex constants, and

$$\omega_m = \alpha \lambda_m \tag{3.50}$$

Writing

$$\omega_m = \omega_{m_r} + \iota \omega_{m_i} \tag{3.51}$$

where ω_{m_r} and ω_{m_i} are the real and imaginary parts of ω_m, gives

$$\tilde{Q}_{k_m}(t) = \tilde{Q}_{k_m}(0)e^{-\omega_{m_r}t}e^{-\iota \omega_{m_i}t} \tag{3.52}$$

Thus, the condition for stability (*i.e.*, the solution does not grow exponentially in time) is

$$\omega_{m_r} \geq 0 \tag{3.53}$$

From (3.44) and (2.12), we have

$$\begin{aligned} \omega_{m_r} &= 0 \\ \omega_{m_i} &= \frac{\sin k\Delta x}{\Delta x}\lambda_m \end{aligned} \tag{3.54}$$

Therefore, the flux quadrature formula (3.27) is stable.

This is shown in Fig. 3.3 for the problem described in Section 2.8. The initial condition is defined by (2.92) with $\epsilon = 0.1$, and the domain $0 < \kappa x < 2\pi$ is discretized into 100 uniform cells where $\kappa = 2\pi L^{-1}$. Periodic boundary conditions are imposed on the left and right boundaries. The initial condition at $\kappa a_o t = 0$ and the solution at $\kappa a_o t = 1.25$ are shown. The profile at $\kappa a_o t = 1.25$ shows close agreement with the exact solution.†

3.8.2 Another Simple Flux Quadrature

We further illustrate the concept of spatial stability by considering a different formula for the flux quadrature

$$F_{i+\frac{1}{2}} = \mathcal{F}(Q_{i+\frac{1}{2}})$$

† However, the agreement degenerates when the shock wave forms (Section 3.8.4).

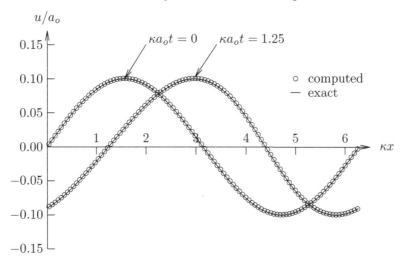

Fig. 3.3. Computed and exact solutions for the flux quadrature (3.27)

$$Q_{i+\frac{1}{2}} = Q_i \tag{3.55}$$

This corresponds to using the Q_i of the cell to the *left side* of the cell face at $x_{i+\frac{1}{2}}$ to determine $F_{i+\frac{1}{2}}$. Proceeding as before, Equation (3.39) is replaced by

$$\frac{dQ_i}{dt} + A_i \frac{(Q_i - Q_{i-1})}{\Delta x} = 0 \tag{3.56}$$

and Equation (3.41) by

$$\frac{d\hat{Q}_k}{dt} + G\hat{Q}_k = 0 \quad \text{for } l = -N+1, \ldots, N \tag{3.57}$$

where

$$G = \alpha A \tag{3.58}$$

and

$$\alpha = \frac{(1 - \cos k\Delta x)}{\Delta x} + \iota \frac{\sin k\Delta x}{\Delta x} \tag{3.59}$$

Following the same analysis used previously,

$$\omega_{m_r} = \frac{(1 - \cos k\Delta x)}{\Delta x} \lambda_m \tag{3.60}$$

$$\omega_{m_i} = \frac{\sin k\Delta x}{\Delta x} \lambda_m \tag{3.61}$$

for $m = 1, 2, 3$. Since $1 - \cos k\Delta x \geq 0$ for all k, $\omega_{m_r} \geq 0$ requires

$$\lambda_1 \geq 0$$

$$\lambda_2 \geq 0$$
$$\lambda_3 \geq 0 \tag{3.62}$$

which implies

$$u \geq a \tag{3.63}$$

Thus, the method in (3.55) is stable only for fluid traveling to the right at or above the speed of sound. This is consistent with flow physics. The formula (3.55) for the flux $F_{i+\frac{1}{2}}$ uses information only from the cell to the left of the face at $i + \frac{1}{2}$. This is physically sensible only if all three characteristics defined by (2.35) have positive slope as illustrated in Fig. 3.4, *i.e.*, the flow at $x_{i+\frac{1}{2}}$ only depends on conditions at $x < x_{i+\frac{1}{2}}$. The algorithm (3.55) is of limited usefulness, however, due to the condition (3.63). If it were to be employed, it would be necessary to test for satisfaction of this condition at every timestep and every face and employ some other method if this condition were not met.

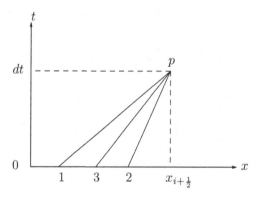

Fig. 3.4. The three characteristics intersecting $x_{i+\frac{1}{2}}$ are shown with positive slopes

An example of the formation of the instability is shown in Fig. 3.5 for the problem described in Section 2.8. The initial condition is defined by (2.92) with $\epsilon = 0.1$, and the domain $0 < \kappa x < 2\pi$ is discretized into 100 uniform cells. Periodic boundary conditions are imposed on the left and right boundaries. The initial condition at $\kappa a_o t = 0$ and the solution at $\kappa a_o t = 1.25$ are shown. A numerical instability is apparent at $\kappa a_o t = 1.25$ wherein the computed velocity displays a pointwise fluctuation. The instability is expected as the condition (3.63) is not met.

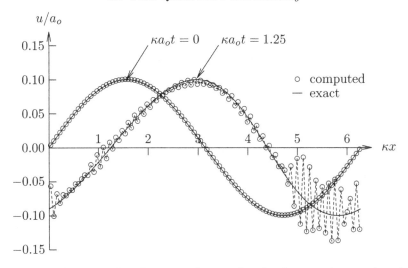

Fig. 3.5. Computed and exact solutions for the flux quadrature (3.55)

3.8.3 Numerical Domain of Dependence

It is useful to consider why the algorithm (3.27) does not exhibit a constraint similar to (3.63). The explanation is again based on flow physics. Regardless of the magnitude of u relative to a and the sign of u, the algorithm (3.27) incorporates flow information corresponding to all three flow characteristics in (2.35). Indeed, algorithm (3.27) incorporates *more* information than necessary under certain circumstances. However, an excess of information regarding the flowfield is not destabilizing, whereas an absence of relevant flow information is destabilizing. In other words, *the numerical domain of dependence must include the physical domain of dependence.*

In summary, we have observed† that the flow physics must be fully respected by the spatial flux quadrature. This requirement is independent of whatever algorithm is chosen for the temporal integration. This is a principle reason for the development of flux algorithms that are strongly motivated by flow physics. We shall discuss some of these algorithms in Chapters 5 and 6.

---

† In a heuristic, and admitedly nonrigorous, manner.

3.8.4 Shock Waves and Weak Solutions

It must be noted, however, that the simple flux algorithm

$$F_{i+\frac{1}{2}} = \mathcal{F}(Q_{i+\frac{1}{2}})$$
$$Q_{i+\frac{1}{2}} = \tfrac{1}{2}(Q_i + Q_{i+1})$$

is not adequate in general, despite satisfying the stability condition (3.53). As noted in Section 2.8, the initial sinusoidal disturbance forms a train of shock waves in a finite time. Figure 3.6 shows the computed and exact profiles for the velocity for the same conditions as Fig. 3.3 at the time of initial formation of the shock wave in (2.97). The computed profile displays severe oscillations immediately downstream of the shock wave (*i.e.*, as viewed by an observer traveling with the shock wave). These oscillations are numerical in origin and represent a fundamental shortcoming of the flux algorithm, namely, the assumption that $\mathcal{Q}$ is a continuous function in x. The appearance of the oscillations is analogous to the Gibbs phenomenon in the Fourier series representation of a discontinuous function (see, for example, Greenberg, 1998). A solution to this problem is presented in Chapter 8.

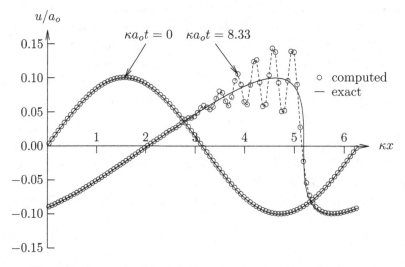

Fig. 3.6. Example of instability associated with a shock wave

3.9 Stability

The selection of a physically consistent spatial flux quadrature is not sufficient to ensure the stability of the discrete approximation (3.13):

$$
\begin{aligned}
Q_i^{n+1} = Q_i^n \quad & - \quad \frac{\Delta t}{\Delta x} \left\{ \theta \left[F(Q_{i+\frac{1}{2}}^{n+1}) - F(Q_{i-\frac{1}{2}}^{n+1}) \right] \right\} \\
& - \quad \frac{\Delta t}{\Delta x} \left\{ (1 - \theta) \left[F(Q_{i+\frac{1}{2}}^{n}) - F(Q_{i-\frac{1}{2}}^{n}) \right] \right\}
\end{aligned}
$$

as we shall now illustrate.

3.9.1 A Simple Flux Quadrature

We assume the simple flux quadrature (3.27),

$$
Q_{i+\frac{1}{2}} = \tfrac{1}{2} \left(Q_{i+1} + Q_i \right)
$$

This algorithm was shown to be stable for the semi-discrete stability analysis
(Section 3.8.1).

The stability analysis of the discrete approximation is similar to the stability analysis for the semi-discrete approximation presented in Section 3.8.
The flow is assumed periodic in x,

$$
Q_1 = Q_M
$$

where M is odd and $M = 2N + 1$. Defining $Q(x, t^n)$ to be a semi-discrete
vector function that interpolates Q_i^n, the Fourier series for $Q(x, t^n)$ is

$$
Q(x, t^n) = \sum_{l=-N+1}^{l=N} \hat{Q}_k^n e^{\iota k x}
$$

where $\iota = \sqrt{-1}$ and the wavenumber k depends on l according to (3.31).
The Fourier coefficients $\hat{Q}_k^n$ are complex vectors whose subscript k indicates
an ordering with respect to the summation index l through (3.31) and the
superscript n indicates a discrete time level t^n. Given the values of Q_i at
some time t^n, the Fourier coefficients $\hat{Q}_k^n$ at t^n are obtained from

$$
\hat{Q}_k^n = \frac{1}{2N} \sum_{i=1}^{i=2N} Q_i(t^n) e^{-\iota k x_i}
$$

The flux vector $F_{i+\frac{1}{2}}$ is expanded in a Taylor series about Q_i,

$$
F_{i+\frac{1}{2}} = F(Q_{i+\frac{1}{2}}) = F(Q_i) + A_i \left(Q_{i+\frac{1}{2}} - Q_i \right) + \mathcal{O} \left((\Delta Q)^2 \right)
$$

where $A = \partial F/\partial Q$ is assumed constant. Thus, (3.13) is approximated by

$$Q_i^{n+1} = Q_i^n \quad - \quad \frac{\Delta t}{\Delta x} A \left[\theta \left(Q_{i+\frac{1}{2}}^{n+1} - Q_{i-\frac{1}{2}}^{n+1} \right) \right]$$
$$- \quad \frac{\Delta t}{\Delta x} A \left[(1 - \theta) \left(Q_{i+\frac{1}{2}}^n - Q_{i-\frac{1}{2}}^n \right) \right]$$

which yields

$$Q_i^{n+1} = Q_i^n \quad - \quad \frac{\Delta t}{2\Delta x} A \left[\theta \left(Q_{i+1}^{n+1} - Q_{i-1}^{n+1} \right) \right]$$
$$- \quad \frac{\Delta t}{2\Delta x} A \left[(1 - \theta) \left(Q_{i+1}^n - Q_{i-1}^n \right) \right]$$

which is a discrete analog of (3.39). Substitute for Q_i from the (interpolating) Fourier series and note that the coefficients of the individual Fourier terms $e^{\iota k x_i}$ must vanish:

$$[I + \Delta t \theta \alpha A] \hat{Q}_k^{n+1} = [I - \Delta t(1 - \theta)\alpha A] \hat{Q}_k^n \tag{3.64}$$

where I is the identity matrix† and

$$\alpha = \frac{\iota \sin k \Delta x}{\Delta x}$$

which is the same as (3.44). From (2.15),

$$A = T \Lambda T^{-1}$$

where Λ is the diagonal matrix of eigenvalues λ_i defined in (2.12). Using (3.46),

$$\tilde{Q}_k = T^{-1} \hat{Q}_k$$

Then (3.64) can be rewritten as

$$[T + \Delta t \, \theta \, \alpha \, T\Lambda] \, \tilde{Q}_k^{n+1} = [T - \Delta t(1 - \theta) \, \alpha \, T\Lambda] \, \tilde{Q}_k^n$$

Multiplying by T^{-1},

$$[I + \Delta t \, \theta \, \alpha \, \Lambda] \, \tilde{Q}_k^{n+1} = [I - \Delta t(1 - \theta) \, \alpha \, \Lambda] \, \tilde{Q}_k^n$$

and therefore

$$\tilde{Q}_k^{n+1} = G \tilde{Q}_k^n \tag{3.65}$$

where

$$G = [I + \Delta t \, \theta \, \alpha \, \Lambda]^{-1} [I - \Delta t(1 - \theta) \, \alpha \, \Lambda] \tag{3.66}$$

---

† The identity matrix has 1's on the diagonal and 0's elsewhere.

It is straightforward to show that

$$G = \left\{ \begin{array}{ccc} \gamma_1 & 0 & 0 \\ 0 & \gamma_2 & 0 \\ 0 & 0 & \gamma_3 \end{array} \right\} \tag{3.67}$$

where γ_i are the eigenvalues of G given by

$$\gamma_i = [1 - \Delta t(1-\theta)\,\alpha\,\lambda_i]\,[1 + \Delta t\,\theta\,\alpha\,\lambda_i]^{-1} \quad \text{for } i = 1, 2, 3 \tag{3.68}$$

The solution for $\tilde{Q}_k^{n+1}$ is therefore

$$\tilde{Q}_k^{n+1} = G^n \tilde{Q}_k^1 \tag{3.69}$$

where $\tilde{Q}_k^1$ is the initial condition and

$$G^n = \underbrace{GG \ldots G}_{n \text{ times}} \tag{3.70}$$

Intuitively, $\tilde{Q}_k^{n+1}$ must remain bounded (in the absence of a physically sensible instability) for any arbitrary initial condition,

$$|\tilde{Q}_k^{n+1}| \leq C |\tilde{Q}_k^1| \quad \text{for all } k \tag{3.71}$$

where $|x|$ indicates the norm[3] of the vector x:

$$|x| = \left(\sum_{i=1}^{i=n} |x_i|^2 \right)^{1/2} \tag{3.72}$$

and C is a constant independent of k. Thus

$$|G^n \tilde{Q}_k^1| \leq C |\tilde{Q}_k^1| \quad \text{for all } k$$

or

$$\max_{k, x \neq 0} \frac{|G^n x|}{|x|} \leq C \tag{3.73}$$

where the maximum is taken over all possible k and all nonzero complex vectors x. The norm $||G||$ of a matrix G is defined as[4]

$$||G|| = \max_{k, x \neq 0} \frac{|Gx|}{|x|} \tag{3.74}$$

and thus (3.73) is equivalent to

$$||G^n|| \leq C \tag{3.75}$$

This is the condition for stability (Richtmyer and Morton, 1967), which may be more precisely stated as

$$G^n \text{ is uniformly bounded} \atop \text{or, equivalently, } \|G^n\| \leq C \quad \text{for} \quad \left\{ \begin{array}{l} 0 < \Delta t < \tau \\ 0 < n\Delta t < T \\ \text{all } k \end{array} \right\} \tag{3.76}$$

for some positive τ and T.

Equation (3.76) is somewhat cumbersome to analyze due to the presence of the power n. We therefore seek a simpler form. It is straightforward to show that, for any matrix G,

$$\|G\| \geq \mathcal{R}(G) \tag{3.77}$$

where $\mathcal{R}(G)$ is the *spectral radius* of G defined by

$$\mathcal{R}(G) = \max_i |\gamma_i| \tag{3.78}$$

where γ_i are the eigenvalues of G. Additionally, it can be shown (Exercise 3.9) that

$$\mathcal{R}^n \leq \|G^n\| \leq \|G\|^n \tag{3.79}$$

Then, a *necessary* condition for stability is

$$\mathcal{R}^n \leq C_2 \quad \text{for} \quad \left\{ \begin{array}{l} 0 < \Delta t < \tau \\ 0 < n\Delta t < T \\ \text{all } k \end{array} \right\} \tag{3.80}$$

for some positive τ, where C_2 is a constant independent of k. Without loss of generality† we may take $C_2 \geq 1$. Then,

$$\mathcal{R} \leq C_2^{\frac{1}{n}} = C_2^{\frac{\Delta t}{T}}$$

Now

$$C_2^{\frac{\Delta t}{T}} = \exp(\tfrac{\Delta t}{T} \log C_2) = 1 + \mathcal{O}(\Delta t)$$

Thus $C_2^{\frac{\Delta t}{T}}$ is bounded by $1 + C_3 \Delta t$ for $0 < \Delta t < \tau$ and hence (3.80) becomes

$$|\gamma_i| \leq 1 + \mathcal{O}(\Delta t) \quad \text{for} \quad \left\{ \begin{array}{l} 0 < \Delta t < \tau \\ 0 < n\Delta t < T \\ \text{all } k \end{array} \right\} \tag{3.81}$$

---

† If in fact $C' < 1$, we could redefine C_2 to be equal to some value greater than 1 and the necessary condition (3.80) would still hold.

for all eigenvalues γ_i of G. This is the *von Neumann condition for stability*. The term $\mathcal{O}(\Delta t)$ permits exponential growth where physically sensible (Richtmyer and Morton, 1967); otherwise, the term $\mathcal{O}(\Delta t)$ is omitted.

If G is a normal matrix‡, then (3.79) becomes (Exercise 3.8)

$$\mathcal{R}^n = ||G^n|| = ||G||^n \tag{3.82}$$

and the von Neumann condition is then clearly sufficient for stability.

It is evident from (3.67) that G is a normal matrix, and therefore the necessary and sufficient condition for stability is

$$\left| \frac{1 - \alpha\,\Delta t\,(1-\theta)\lambda_i}{1 + \alpha\,\Delta t\,\theta\lambda_i} \right| \leq 1 \tag{3.83}$$

assuming there are no physically sensible instabilities. Define

$$\alpha = \iota\beta \quad \text{where} \quad \beta = \frac{\sin k\Delta x}{\Delta x} \tag{3.84}$$

Then (3.83) becomes

$$(\beta\,\Delta t\,\lambda_i)^2\,(1-2\theta) \leq 0 \tag{3.85}$$

for all i. The condition for stability is

$$\theta \geq \tfrac{1}{2} \tag{3.86}$$

Thus, for the flux quadrature

$$Q_{i+\frac{1}{2}} = \tfrac{1}{2}\,(Q_i + Q_{i+1})$$

Euler explicit ($\theta = 0$) is *unconditionally unstable*†, while Euler implicit ($\theta = 1$) and Crank-Nicholson ($\theta = \frac{1}{2}$) are unconditionally stable. The instability for Euler explicit is shown in Fig. 3.7 for the problem described in Section 2.8 with $\epsilon = 0.01$, the domain $0 < \kappa x < 2\pi$ discretized into 100 uniform cells, and a time step $\Delta t = \Delta t_{CFL}$, where Δt_{CFL} is defined in (3.94).

It is therefore evident that stability for the semi-discrete approximation is not a sufficient condition for stability of the discrete approximation.

---

‡ A normal matrix is an $n \times n$ matrix with n orthogonal eigenvectors. It can be shown (Franklin, 1968) that a normal matrix commutes with its adjoint, *i.e.*, $GG^* = G^*G$, where G^* is the complex conjugate transpose of G.

† See, for example, Anderson *et al.* (1984).

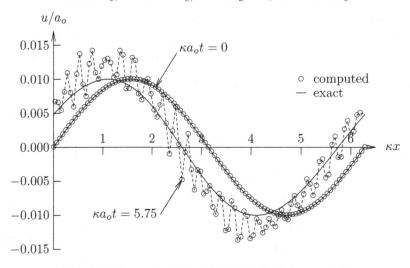

Fig. 3.7. Example of instability with Euler explicit

3.9.2 Another Simple Flux Quadrature

We consider the simple flux quadrature (3.55),

$$Q_{i+\frac{1}{2}} = Q_i$$

This algorithm was shown to be stable for the semi-discrete analysis (Section 3.8.2) provided

$$\lambda_i \geq 0 \quad \text{for} \quad i = 1, 2, 3 \tag{3.87}$$

By a similar analysis,

$$[I + \Delta t \theta \alpha A] \, \hat{Q}_k^{n+1} = [I - \Delta t(1 - \theta)\alpha A] \, \hat{Q}_k^n \tag{3.88}$$

where

$$\alpha = a + \iota b \tag{3.89}$$

where

$$
\begin{aligned}
a &= \frac{(1 - \cos k\Delta x)}{\Delta x} \\
b &= \frac{\sin k\Delta x}{\Delta x}
\end{aligned}
\tag{3.90}
$$

The necessary and sufficient condition for stability is again (3.83) where α is defined by (3.89) and (3.90). The result is

$$\left(\frac{\Delta t \, \lambda_i}{\Delta x}\right) \left[\left(\frac{\Delta t \, \lambda_i}{\Delta x}\right)(1 - 2\theta) - 1\right] \leq 0 \tag{3.91}$$

for $i = 1, 2, 3$.

For Euler explicit ($\theta = 0$), the algorithm is *conditionally stable* with the restrictions

$$\lambda_i \geq 0 \ \text{ and } \ \Delta t \leq \frac{\Delta x}{\lambda_i}$$

for $i = 1, 2, 3$. Thus,

$$u \ \geq \ a \tag{3.92}$$
$$\Delta t \ \leq \ \Delta t_{CFL} \tag{3.93}$$

where

$$\Delta t_{CFL} = \min_i \frac{\Delta x}{\lambda_i} \tag{3.94}$$

More generally,

$$\Delta t_{CFL} = \min_i \frac{\Delta x}{|\lambda_i|} \tag{3.95}$$

The second restriction in (3.93) is the *Courant-Friedrichs-Lewy* (or *CFL*) condition. The *Courant number* $\mathcal{C}$ is defined as

$$\mathcal{C} = \frac{\Delta t}{\Delta t_{CFL}} \tag{3.96}$$

and the second restriction can be rewritten as

$$\mathcal{C} \leq 1 \tag{3.97}$$

Note that the first restriction in (3.93) is identical to the condition obtained for the semi-discrete stability analysis in (3.63).

For Crank Nicholson ($\theta = \frac{1}{2}$), the algorithm is *conditionally stable* with the restriction

$$\lambda_i \geq 0 \tag{3.98}$$

This, too, is identical to (3.63).

For Euler implicit ($\theta = 1$), the stability condition is

$$\Delta t \geq \max_i \left(-\frac{\Delta x}{\lambda_i} \right) \tag{3.99}$$

For $\lambda_i \geq 0$, the algorithm is unconditionally stable. If one or more λ_i is negative, then the algorithm is conditionally stable with a minimum allowable timestep.

3.10 Convergence

In the context of a finite volume algorithm, a numerical solution Q_i^n converges to the exact solution $Q_i^{n,e}$ iff†

$$\lim_{\substack{\Delta t \to 0 \\ \Delta x \to 0}} \|Q_i^{n,e} - Q_i^n\| = 0 \qquad (3.100)$$

at a fixed value t where $Q_i^{n,e}$ is the volume average of the exact solution over cell i:

$$Q_i^{n,e} = \frac{1}{V_i} \int_{V_i} Q \, dx \, dy \qquad (3.101)$$

We employ the root-mean-square for the norm in (3.100) as

$$\|Q_i^{n,e} - Q_i^n\| = \left\{ \frac{1}{M} \sum_{\rho, \rho u, \rho e} \mathcal{Q}_\infty^{-2} \sum_{i=1}^{i=M} (Q_i^{n,e} - Q_i^n)^2 \right\}^{1/2} \qquad (3.102)$$

where the first $\sum$ implies summation over the three conservative variables and $\mathcal{Q}_\infty = (\rho_o, \rho_o a_o, \rho_o a_o^2)^{\mathrm{T}}$. For a spatially and temporally second-order accurate algorithm,

$$Q_i^{n+1} = \mathcal{Q}_i^{n+1} + \mathcal{O}(\Delta t^2, \Delta x^2)$$

and therefore

$$\|Q_i^{n,e} - Q_i^n\| = \mathcal{O}(\Delta t^2, \Delta x^2)$$

The results for the simple flux quadrature (3.27) are presented in Fig. 3.8 for the problem described in Section 2.8. The initial condition is defined by (2.92) with $\epsilon = 0.1$, and the domain is $0 < \kappa x < 2\pi$. The norm is evaluated at $\kappa a_o t = 7$, which is prior to the shock formation, and $\Delta t \propto \Delta x$. The exact solution $\mathcal{Q}_i^n$ is obtained from (2.85) to (2.88). It is evident that the computed solution converges quadratically† to the exact solution for sufficiently small Δt.

The Equivalence Theorem of Lax and Richtmyer (1956) provides an important insight to the issue of convergence, as it is usually difficult to demonstrate the convergence of the numerical solution except for simple problems

---

† We use the double vertical brace notation $\| \cdot \|$ to denote the norm of the composite of the solutions for ρ, ρu, and ρe as defined in (3.102). This is not be confused with the matrix norm introduced in Section 3.9.

† The quadratic line in Fig. 3.8 is

$$\|Q_i^{n,e} - Q_i^n\|_{\kappa a_o \Delta t} = \left(\frac{\kappa a_o \Delta t}{10^{-3}} \right)^2 \|Q_i^{n,e} - Q_i^n\|_{\kappa a_o \Delta t = 10^{-3}}$$

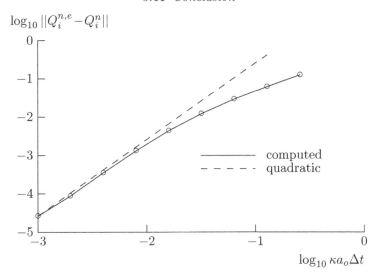

Fig. 3.8. Convergence study

whose exact solutions are known. The theorem applies to linear systems (with properly posed initial conditions) discretized by a consistent finite difference approximation. Under these conditions, the theorem states that the stability of the finite difference approximation is a necessary and sufficient condition for convergence. In other words, consistency and stability imply convergence.

3.11 Conclusion

In this chapter, we have introduced a family of discrete approximations (3.13) to the Euler equations for the purpose of illustrating the concepts of accuracy, consistency, convergence, and stability. This particular discretization incorporates both unconditionally and conditionally stable methods, and unstable methods. It is by no means a full catalog of discretization schemes, and a significant number of other algorithms have been developed.[5]

Exercises

3.1 Show that a) $\sin x = \mathcal{O}(x)$ as $x \to 0$, b) $\cos x = \mathcal{O}(x - \pi/2)$ as $x \to \pi/2$, and c) $x \log x = o(1)$ as $x \to 0$

SOLUTION

a) By L'Hospital's rule,

$$\lim_{x \to 0} \frac{\sin x}{x} = \lim_{x \to 0} \frac{\cos x}{1} = 1$$

Thus, $\sin x = \mathcal{O}(x)$ as $x \to 0$.

b) By L'Hospital's rule,

$$\lim_{x \to \pi/2} \frac{\cos x}{(x - \pi/2)} = \lim_{x \to \pi/2} -\frac{\sin x}{1} = -1$$

Thus, $\cos x = \mathcal{O}(x - \pi/2)$ as $x \to \pi/2$.

c) By L'Hospital's rule,

$$\lim_{x \to 0} x \log x = \lim_{x \to 0} \frac{\log x}{x^{-1}} = \lim_{x \to 0} -\frac{x^{-1}}{x^{-2}} = \lim_{x \to 0} -x = 0$$

Thus, $x \log x = o(1)$ as $x \to 0$.

3.2 Repeat the derivation in Section 3.8.2 with $Q_{i+\frac{1}{2}} = Q_{i+1}$ and show the method is stable only if $u \leq -a$. Discuss the physical significance.

3.3 Find an explicit expression for the $\mathcal{O}(\Delta Q)^2$ terms in (3.34).

SOLUTION

The Taylor series expansion of a function $f(x_1 + \Delta x_1, x_2 + \Delta x_2, x_3 + \Delta x_3)$ about (x_1, x_2, x_3) can be expressed as

$$f(x_1 + \Delta x_1, x_2 + \Delta x_2, x_3 + \Delta x_3) = f(x_1, x_2, x_3) +$$

$$\begin{bmatrix} f_{x_1} & f_{x_2} & f_{x_3} \end{bmatrix} \begin{bmatrix} \Delta x_1 \\ \Delta x_2 \\ \Delta x_3 \end{bmatrix} +$$

$$\frac{1}{2} \begin{bmatrix} \Delta x_1 & \Delta x_2 & \Delta x_3 \end{bmatrix} \mathcal{H} \begin{bmatrix} \Delta x_1 \\ \Delta x_2 \\ \Delta x_3 \end{bmatrix}$$

where $\mathcal{H}$ is the Hessian matrix

$$\mathcal{H} = \begin{Bmatrix} f_{x_1 x_1} & f_{x_1 x_2} & f_{x_1 x_3} \\ f_{x_2 x_1} & f_{x_2 x_2} & f_{x_2 x_3} \\ f_{x_3 x_1} & f_{x_3 x_2} & f_{x_3 x_3} \end{Bmatrix}$$

and $f_{x_1} = \partial f / \partial x_1$, etc. Denote $\mathcal{H}_1$, $\mathcal{H}_2$, and $\mathcal{H}_3$ as the Hessian matrices for the conservation of mass, momentum, and energy. Then,

$$\mathcal{H}_1 = \begin{Bmatrix} 0 & 0 & 0 \\ 0 & 0 & 0 \\ 0 & 0 & 0 \end{Bmatrix}$$

$$\mathcal{H}_2 = \begin{Bmatrix} -(\gamma-3)\frac{Q_2^2}{Q_1^3} & (\gamma-3)\frac{Q_2}{Q_1^2} & 0 \\ (\gamma-3)\frac{Q_2}{Q_1^2} & -(\gamma-3)\frac{1}{Q_1} & 0 \\ 0 & 0 & 0 \end{Bmatrix}$$

$$\mathcal{H}_3 = \begin{Bmatrix} 2\gamma\frac{Q_2 Q_3}{Q_1^3} - 3(\gamma-1)\frac{Q_2^3}{Q_1^4} & -\gamma\frac{Q_3}{Q_1^2} + 3(\gamma-1)\frac{Q_2^2}{Q_1^3} & -\gamma\frac{Q_2}{Q_1^2} \\ -\gamma\frac{Q_3}{Q_1^2} + 3(\gamma-1)\frac{Q_2^2}{Q_1^3} & -3(\gamma-1)\frac{Q_2}{Q_1^2} & \frac{\gamma}{Q_1} \\ -\gamma\frac{Q_2}{Q_1^2} & \frac{\gamma}{Q_1} & 0 \end{Bmatrix}$$

3.4 In deriving (3.41) and (3.64), it was stated that the coefficients of the individual Fourier terms $e^{\iota k x_i}$ must vanish. Prove this statement.

3.5   Prove that (3.62) implies (3.63).

Equations (3.62) are

$$
\begin{aligned}
u &\geq 0 \\
u + a &\geq 0 \\
u - a &\geq 0
\end{aligned}
$$

which can be written

$$
\begin{aligned}
u &\geq 0 \\
u &\geq -a \\
u &\geq a
\end{aligned}
$$

The third condition implies the first and second and therefore is the requirement.

3.6   Show that $||G|| \geq \mathcal{R}(G)$, where $\mathcal{R}(G)$ is the spectral radius of G.

3.7   Prove that if $||G|| < 1$, then

$$
\frac{1}{1 + ||G||} \leq ||(1 - G)^{-1}|| \leq \frac{1}{1 - ||G||}
$$

SOLUTION
From the identity

$$
I = (1 - G)(1 - G)^{-1}
$$

we have

$$
||I|| = ||(1 - G)(1 - G)^{-1}||
$$

Now

$$
||(1 - G)(1 - G)^{-1}|| \leq ||(1 - G)|| \, ||(1 - G)^{-1}||
$$

and

$$
||(1 - G)|| \leq (1 + ||G||)
$$

Thus, since $||I|| = 1$,

$$
1 \leq (1 + ||G||) \, ||(1 - G)^{-1}||
$$

and therefore

$$
\frac{1}{1 + ||G||} \leq ||(1 - G)^{-1}||
$$

From

$$
(I - G)(I - G)^{-1} = I
$$

we obtain

$$
(I - G)^{-1} = I + G(I - G)^{-1}
$$

and thus

$$
||(I - G)^{-1}|| \leq 1 + ||G|| \, ||(I - G)^{-1}||
$$

Since $||G|| \leq 1$,

$$
||(I - G)^{-1}|| \leq \frac{1}{1 - ||G||}
$$

(Isaacson and Keller, 1966).

3.8   Show that if G is a normal matrix, then $\mathcal{R}^n = ||G^n|| = ||G||^n$, where $\mathcal{R}$ is the spectral radius of G.

3.9   Show that, in general, $\mathcal{R}^n \leq ||G^n|| \leq ||G||^n$, where $\mathcal{R}$ is the spectral radius of G.

SOLUTION

The proof for $n = 1$ is Exercise 3.6. Let $n = 2$ and let $\mathcal{V}$ denote the set of right eigenvectors $e^i = (e_1^i, e_2^i, \ldots, e_n^i)^T$ of G with eigenvalues λ_i. Then

$$
\begin{aligned}
||G^2|| &= \max_{x \neq 0} \frac{|GGx|}{|x|} \\
&= \max_{x \neq 0} \frac{|GGx|}{|Gx|} \frac{|Gx|}{|x|} \\
&\geq \max_{x \in \mathcal{V}} \frac{|GGx|}{|Gx|} \frac{|Gx|}{|x|}
\end{aligned}
$$

where the inequality arises since the set of eigenvectors $\mathcal{V}$ is a subset of all nonzero vectors x. Furthermore,

$$
\begin{aligned}
\max_{x \in \mathcal{V}} \frac{|GGx|}{|Gx|} \frac{|Gx|}{|x|} &= \max_i \frac{|G\lambda_i e^i|}{|\lambda_i e^i|} \frac{|\lambda_i e^i|}{|e^i|} \\
&= \max_i |\lambda_i| \frac{|Ge^i|}{|\lambda_i| |e^i|} \frac{|\lambda_i| |e^i|}{|e^i|} \\
&= \max_i |\lambda_i|^2 \\
&= \left(\max_i |\lambda_i| \right)^2 \\
&= \mathcal{R}^2
\end{aligned}
$$

The proof for the general case follows by induction. The second inequality follows from the matrix norm property $||AB|| \leq ||A|| \, ||B||$, where A and B are matrices (see the Endnotes).

3.10   Derive the stability condition (3.86) from (3.83).

4

Reconstruction

There is an extensive mathematical literature devoted to theorems about what sort of functions can be well approximated by which interpolating functions. These theorems are, alas, almost completely useless in day-to-day work: if we know enough about our function to apply a theorem of any power, we are usually not in the pitiful state of having to interpolate on a table of its values!

William Press et al. (1986)

4.1 Introduction

The semi-discrete form of the Euler equations (3.9) is

$$\frac{dQ_i}{dt} + \frac{\left(F_{i+\frac{1}{2}} - F_{i-\frac{1}{2}}\right)}{\Delta x} = 0 \tag{4.1}$$

where $Q_i(t)$ is the cell averaged vector of dependent variables,

$$Q_i(t) = \frac{1}{V_i} \int_{V_i} Q \, dx dy$$

$F_{i\pm\frac{1}{2}}$ is the spatial flux quadrature

$$F_{i\pm\frac{1}{2}} = \frac{1}{A_{i\pm\frac{1}{2}}} \int_{x_{i\pm\frac{1}{2}}} \mathcal{F} dy$$

and $A_{i\pm\frac{1}{2}} = \Delta y$ is the surface area of the face at $x_{i\pm\frac{1}{2}}$. For a one-dimensional flow,

$$F_{i\pm\frac{1}{2}} = \mathcal{F}_{i\pm\frac{1}{2}}$$

where $\mathcal{F}(\mathcal{Q})$ is given by (2.5).

The discretization of the domain (Section 3.3) and introduction of the volume averaged vector $Q_i(t)$ result in a loss of information regarding $\mathcal{Q}(x,t)$. Consider, for example, the periodic function $\mathcal{Q}(x) = \sin x$ for $0 \leq x \leq 2\pi$. Assume ten cells are employed. The exact function $\mathcal{Q}(x)$ and cell averaged values Q_i are shown in Fig. 4.1. Within a given cell, the cell averaged value is only an approximation of the exact function $\mathcal{Q}$. Of course, the approximation improves as the size of the cell is reduced.

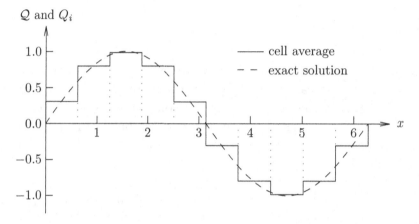

Fig. 4.1. Exact function $\mathcal{Q}(x) = \sin x$ and cell averaged Q_i

The time evolution of Q_i requires, from (4.1), the fluxes $F_{i\pm\frac{1}{2}}$, which, in turn, must be computed using the Q_i in the vicinity of $x_{i\pm\frac{1}{2}}$. Within each cell i, a local approximate *reconstruction* $Q_i(x)$ of the exact function $\mathcal{Q}(x)$ can be formed to compute the fluxes.† Note, however, that a discontinuity may exist at each cell interface, *i.e.*, $Q_i(x_{i+\frac{1}{2}}) \neq Q_{i+1}(x_{i+\frac{1}{2}})$. The algorithm for the fluxes $F_{i\pm\frac{1}{2}}$ must take this discontinuous behavior into consideration.

The simplest reconstruction is $Q_i(x) = Q_i$, which is first-order accurate. This method leads to excessive numerical diffusion, however, and is generally not acceptable. We therefore seek reconstruction methods of higher order accuracy. Three methods are presented in the following sections. Methods for computing $F_{i+\frac{1}{2}}$ are described in Chapters 5 and 6.

---

† Note that Q_i is the cell averaged value in cell i and is therefore a constant. $Q_i(x)$ is a function defined in cell i that is in general not a constant. $Q_i(x)$ is constructed to provide a closer approximation to $\mathcal{Q}(x)$ than is afforded by Q_i.

4.2 Reconstruction Using the Primitive Function

Colella and Woodward (1984) and Harten *et al.* (1987a) developed methods for higher order reconstruction‡. In each cell i, a polynomial is constructed to approximate Q. On the right face of cell i at $x_{i-\frac{1}{2}}$, the polynomial is employed to estimate the value§ of Q denoted by $Q^r_{i-\frac{1}{2}}$, while on the left face at $x_{i+\frac{1}{2}}$, the polynomial is employed to estimate the value of Q denoted by $Q^l_{i+\frac{1}{2}}$ (Fig. 4.2). A similar reconstruction is performed for all cells. For cell i, the flux $F_{i+\frac{1}{2}}$ is then determined using $Q^l_{i+\frac{1}{2}}$ and $Q^r_{i+\frac{1}{2}}$, and similarly the flux $F_{i-\frac{1}{2}}$ is then determined using $Q^l_{i-\frac{1}{2}}$ and $Q^r_{i-\frac{1}{2}}$, This is summarized in Table 4.1.

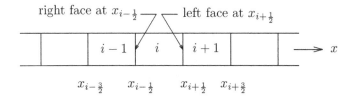

Fig. 4.2. Reconstruction to interior faces of cell i

Table 4.1. *Reconstruction of Q for Computing $F_{i\pm\frac{1}{2}}$*

Quantity	Use reconstruction in cell
$Q^l_{i-\frac{1}{2}}$	$i-1$
$Q^r_{i-\frac{1}{2}}$	i
$Q^l_{i+\frac{1}{2}}$	i
$Q^r_{i+\frac{1}{2}}$	$i+1$

We present a particular form of a reconstruction method that employs the cell averaged values Q_{i-1}, Q_i, and Q_{i+1} to reconstruct the solution in cell i. We define a *primitive function* $I(x)$ according to

$$I(x) = \int_{x_{i-\frac{3}{2}}}^{x} Q \, dx \quad \text{for} \quad x_{i-\frac{3}{2}} \le x \le x_{i+\frac{3}{2}} \tag{4.2}$$

where Δx is assumed constant. Note that $I(x)$ is defined for x within the

---

‡ Earlier, Van Leer (1979) developed a second-order accurate method using piecewise linear reconstruction with Godunov's method.

§ It is also possible to interpolate a different set of variables, *e.g.*, velocity, pressure, and density.

limits of the three cells $i-1, i$, and $i+1$. Then

$$
\begin{aligned}
I(x_{i-\frac{3}{2}}) &= 0 \\
I(x_{i-\frac{1}{2}}) &= \Delta x \, Q_{i-1} \\
I(x_{i+\frac{1}{2}}) &= \Delta x \, (Q_{i-1} + Q_i) \\
I(x_{i+\frac{3}{2}}) &= \Delta x \, (Q_{i-1} + Q_i + Q_{i+1})
\end{aligned} \tag{4.3}
$$

There is a unique third-order polynomial $P(x)$ that interpolates $I(x)$ at the four points $x_{i-\frac{3}{2}}$, $x_{i-\frac{1}{2}}$, $x_{i+\frac{1}{2}}$, and $x_{i+\frac{3}{2}}$. The polynomial may be obtained using Newton's formula (Mathews and Fink, 1998)

$$
\begin{aligned}
P(x) &= a_0 + a_1(x - x_{i-\frac{3}{2}}) + a_2(x - x_{i-\frac{3}{2}})(x - x_{i-\frac{1}{2}}) + \\
&\quad a_3(x - x_{i-\frac{3}{2}})(x - x_{i-\frac{1}{2}})(x - x_{i+\frac{1}{2}})
\end{aligned} \tag{4.4}
$$

The coefficients are defined by

$$
\begin{aligned}
a_0 &= I[x_{i-\frac{3}{2}}] \\
a_1 &= I[x_{i-\frac{3}{2}}, x_{i-\frac{1}{2}}] \\
a_2 &= I[x_{i-\frac{3}{2}}, x_{i-\frac{1}{2}}, x_{i+\frac{1}{2}}] \\
a_3 &= I[x_{i-\frac{3}{2}}, x_{i-\frac{1}{2}}, x_{i+\frac{1}{2}}, x_{i+\frac{3}{2}}]
\end{aligned} \tag{4.5}
$$

where

$$
\begin{aligned}
I[x_{i-\frac{3}{2}}] &= I(x_{i-\frac{3}{2}}) \\
I[x_{i-\frac{3}{2}}, x_{i-\frac{1}{2}}] &= \frac{I[x_{i-\frac{1}{2}}] - I[x_{i-\frac{3}{2}}]}{x_{i-\frac{1}{2}} - x_{i-\frac{3}{2}}} \\
I[x_{i-\frac{3}{2}}, x_{i-\frac{1}{2}}, x_{i+\frac{1}{2}}] &= \frac{I[x_{i-\frac{1}{2}}, x_{i+\frac{1}{2}}] - I[x_{i-\frac{3}{2}}, x_{i-\frac{1}{2}}]}{x_{i+\frac{1}{2}} - x_{i-\frac{3}{2}}} \\
I[x_{i-\frac{3}{2}}, x_{i-\frac{1}{2}}, x_{i+\frac{1}{2}}, x_{i+\frac{3}{2}}] &= \frac{I[x_{i-\frac{1}{2}}, x_{i+\frac{1}{2}}, x_{i+\frac{3}{2}}] - I[x_{i-\frac{3}{2}}, x_{i-\frac{1}{2}}, x_{i+\frac{1}{2}}]}{x_{i+\frac{3}{2}} - x_{i-\frac{3}{2}}}
\end{aligned} \tag{4.6}
$$

Now

$$
I(x) = P(x) + E(x) \quad \text{for} \quad x_{i-\frac{3}{2}} \le x \le x_{i+\frac{3}{2}} \tag{4.7}
$$

where $E(x)$ is the error defined by

$$
E(x) = \frac{(x-x_{i-\frac{3}{2}})(x-x_{i-\frac{1}{2}})(x-x_{i+\frac{1}{2}})(x-x_{i+\frac{3}{2}})}{4!} \frac{d^4 I}{dx^4}\bigg|_{x=\hat{x}} \tag{4.8}
$$

where $\hat{x}$ depends on x and $x_{i-\frac{3}{2}} \le \hat{x} \le x_{i+\frac{3}{2}}$.

The reconstruction function† for $\mathcal{Q}$ in cell i is denoted as $Q(x)$ and is defined by

$$Q_i(x) = \frac{dP}{dx} \quad \text{for} \quad x_{i-\frac{1}{2}} \le x \le x_{i+\frac{1}{2}} \tag{4.9}$$

Thus

$$
\begin{aligned}
Q_i(x) \;=\; & a_1 + a_2 \left[(x-x_{i-\frac{1}{2}}) + (x-x_{i-\frac{3}{2}}) \right] + \\
& a_3 \left[(x-x_{i-\frac{1}{2}})(x-x_{i+\frac{1}{2}}) + (x-x_{i-\frac{3}{2}})(x-x_{i+\frac{1}{2}}) + \right. \\
& \left. (x-x_{i-\frac{3}{2}})(x-x_{i-\frac{1}{2}}) \right]
\end{aligned}
$$

Using (4.3), (4.5), and (4.6),

$$
\begin{aligned}
I[x_{i-\frac{3}{2}}, x_{i-\frac{1}{2}}] &= Q_{i-1} \\
I[x_{i-\frac{1}{2}}, x_{i+\frac{1}{2}}] &= Q_i \\
I[x_{i+\frac{1}{2}}, x_{i+\frac{3}{2}}] &= Q_{i+1} \\
I[x_{i-\frac{3}{2}}, x_{i-\frac{1}{2}}, x_{i+\frac{1}{2}}] &= (2\Delta x)^{-1}\left(Q_i - Q_{i-1}\right) \\
I[x_{i-\frac{1}{2}}, x_{i+\frac{1}{2}}, x_{i+\frac{3}{2}}] &= (2\Delta x)^{-1}\left(Q_{i+1} - Q_i\right) \tag{4.10}
\end{aligned}
$$

Defining

$$\Delta Q_{i+\frac{1}{2}} = Q_{i+1} - Q_i \tag{4.11}$$

the coefficients are

$$
\begin{aligned}
a_1 &= Q_{i-1} \\
a_2 &= (2\Delta x)^{-1}\Delta Q_{i-\frac{1}{2}} \\
a_3 &= \left[6(\Delta x)^2\right]^{-1}\left(\Delta Q_{i+\frac{1}{2}} - \Delta Q_{i-\frac{1}{2}}\right)
\end{aligned}
\tag{4.12}
$$

Define

$$\xi = x - x_i \tag{4.13}$$

and then

$$
\begin{aligned}
Q_i(x) \;=\; & Q_i - \frac{\left(\Delta Q_{i+\frac{1}{2}} - \Delta Q_{i-\frac{1}{2}}\right)}{24} + \frac{\left(\Delta Q_{i+\frac{1}{2}} + \Delta Q_{i-\frac{1}{2}}\right)}{2\Delta x}\,\xi \\
& + \frac{\left(\Delta Q_{i+\frac{1}{2}} - \Delta Q_{i-\frac{1}{2}}\right)}{2(\Delta x)^2}\,\xi^2
\end{aligned}
\tag{4.14}
$$

† Recall that Q_i is the cell averaged value in cell i and is therefore a constant, and $Q_i(x)$ is a function defined in cell i that is in general not a constant.

The domain of dependence of $Q_i(x)$ for cell i is shown in Fig. 4.3. From (4.14) it is evident that $Q_i(x)$ depends on Q_{i-1}, Q_i and Q_{i+1}.

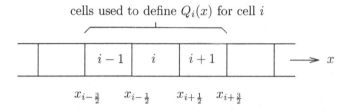

Fig. 4.3. Cells used to define $Q_i(x)$

From (4.2), (4.7), and (4.9),

$$Q_i(x) = \frac{dI}{dx} - \frac{dE}{dx}$$
$$= \mathcal{Q}(x) - \frac{dE}{dx}$$

From (4.8), it is evident that $dE/dx = \mathcal{O}(\Delta x^3)$. Therefore, $Q_i(x)$ is a third-order accurate reconstruction of $\mathcal{Q}$ within cell i.

Equation (4.14) can be applied to define the reconstruction within cell i. By replacing i by $i-1$, the reconstruction within cell $i-1$ is obtained, and similarly for cell $i+1$. Thus,

$$
\begin{aligned}
Q^l_{i+\frac{1}{2}} &= Q_i &+ \tfrac{1}{3}\Delta Q_{i+\frac{1}{2}} + \tfrac{1}{6}\Delta Q_{i-\frac{1}{2}} \\
Q^r_{i+\frac{1}{2}} &= Q_{i+1} - \tfrac{1}{3}\Delta Q_{i+\frac{1}{2}} - \tfrac{1}{6}\Delta Q_{i+\frac{3}{2}} \\
Q^l_{i-\frac{1}{2}} &= Q_{i-1} + \tfrac{1}{3}\Delta Q_{i-\frac{1}{2}} + \tfrac{1}{6}\Delta Q_{i-\frac{3}{2}} \\
Q^r_{i-\frac{1}{2}} &= Q_i &- \tfrac{1}{3}\Delta Q_{i-\frac{1}{2}} - \tfrac{1}{6}\Delta Q_{i+\frac{1}{2}}
\end{aligned}
\qquad (4.15)
$$

Therefore, the computation of the flux contributions to cell i involve cells $i-2$ to $i+2$ as shown in Fig. 4.4.

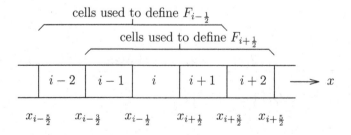

Fig. 4.4. Dependence of $F_{i-\frac{1}{2}}$ and $F_{i+\frac{1}{2}}$

An example of the reconstruction of a scalar function is presented in Fig. 4.5. The exact function is $\mathcal{Q}(x) = \sin x$ and it is assumed periodic. Ten cells are employed. The function is reconstructed within each cell i according to (4.14), with the symbols $\triangleleft$ and $\triangleright$ indicating the values of $Q^r_{i-\frac{1}{2}}$ and $Q^l_{i+\frac{1}{2}}$, respectively, for cell i. It is evident that the reconstructed function $Q_i(x)$ is a close approximation to the exact function $\mathcal{Q}(x)$ within each cell. There is a small jump in $Q_i(x)$ evident at the face of each cell. By denoting

$$\delta Q_{i+\frac{1}{2}} = Q^l_{i+\frac{1}{2}} - Q^r_{i+\frac{1}{2}} \tag{4.16}$$

we have

$$\delta Q_{i+\frac{1}{2}} = \tfrac{1}{6}\left(\Delta Q_{i+\frac{3}{2}} - 2\Delta Q_{i+\frac{1}{2}} + \Delta Q_{i-\frac{1}{2}}\right)$$

and using (4.11),

$$\delta Q_{i+\frac{1}{2}} = \tfrac{1}{6}\left(Q_{i+2} - 3Q_{i+1} + 3Q_i - Q_{i-1}\right) \tag{4.17}$$

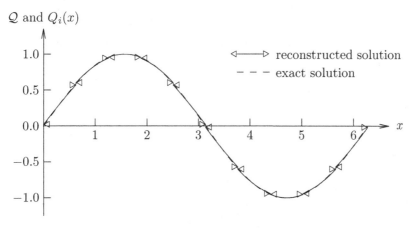

Fig. 4.5. Reconstruction of sine using ten cells

4.3 No New Extrema

A problem arises with the reconstruction (4.14) when the function $\mathcal{Q}$ exhibits a discontinuity as illustrated in Fig. 4.6. The exact function is

$$\mathcal{Q}(x) = \begin{cases} -1 & \text{for } x \leq \pi \\ +1 & \text{for } \pi < x \end{cases} \tag{4.18}$$

Ten cells are employed, with the discontinuity coinciding with the boundary separating the fifth and sixth cells. The function is reconstructed within each cell i according to (4.14), with the symbols $\triangleleft$ and $\triangleright$ indicating the values of $Q^r_{i-\frac{1}{2}}$ and $Q^l_{i+\frac{1}{2}}$, respectively, for cell i.

It is evident that the reconstruction (4.14) is a poor approximation of the exact function Q within the fifth and sixth cells. The reconstruction has introduced *fictitious local extrema* at $x_{4\frac{1}{2}}$ and $x_{6\frac{1}{2}}$. The jump $\delta Q_{5\frac{1}{2}}$ at the discontinuity $(x = \pi)$ is incorrect. Using (4.17),

$$\delta Q_{5\frac{1}{2}} = \tfrac{1}{3}\left(Q_L - Q_R\right)$$

noting that $Q_i = -1$ for $i = 1$ to 5 and $Q_i = +1$ for $i = 6$ to 10. Finally, the reconstruction yields a variation of $Q_i(x)$ with x in the fifth and sixth cells while the exact function Q is a constant.

The source of the problem in Fig. 4.6 can be identified using Fig. 4.3. When cell i is immediately to the left of the discontinuity, the reconstruction algorithm (4.14) employs data from both the left and the right of the discontinuity. A correct reconstruction would only employ data from the left. Similarly, when cell i is immediately to the right of the discontiuity, the reconstruction algorithm (4.14) employs data from both the left and the right of the discontinuity. A correct reconstruction would only employ data from the right.

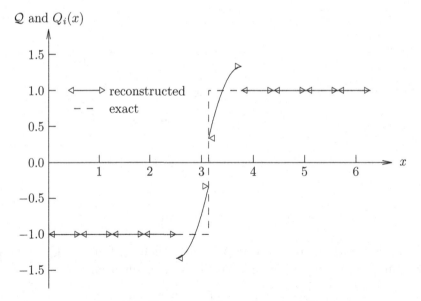

Fig. 4.6. Reconstruction using ten cells

The criteria for developing reconstruction algorithms must therefore be expanded. It is clearly insufficient to simply require a given formal order of accuracy of reconstruction. This approach assumes that the exact function Q possesses continuous derivatives up to some order depending on the order of reconstruction. This assumption is not valid when Q manifests a discontinuity (*e.g.*, a shock or contact surface in gas dynamics).

In principle, the reconstruction algorithm could be revised to incorporate an appropriate directional bias in the vicinity of a discontinuity. However, in solving the Euler equations, the exact function Q is not known, and therefore detection of the discontinuity is considerably more difficult.

A straightforward approach is to avoid the formation of new local extrema relative to the cell averaged solutions in some neighborhood of cell i. This approach is denoted *No New Extrema* and may be summarized as follows:

> *No New Extrema (NNE):* The reconstructed left and right states $Q^l_{i+\frac{1}{2}}$ and $Q^r_{i-\frac{1}{2}}$ shall not introduce any new extrema relative to Q_{i-a}, Q_{i-a+1}, ..., Q_i, ..., Q_{i+b-1}, Q_{i+b}, where $a \geq 0$ and $b \geq 0$.

The constants a and b define the neighborhood of cells surrounding (and including) cell i that are utilized in the determination of the local extrema (local maximum and minimum).† The modification to a reconstruction algorithm to incorporate this principle is denoted a *limiter*. An example is the Modified Upwind Scheme for Conservation Laws (MUSCL) described in the following section. It should be emphasized, however, that the NNE principle does not guarantee that a left or right state is correctly reconstructed at a discontinuity in Q. An counterexample is presented in Fig. 4.22 later.

4.4 Modified Upwind Scheme for Conservation Laws

The reconstruction function (4.14) and face values (4.15) require modification in the presence of discontinuities in Q. We may restrict our attention to the face values since they determine the flux at the cell face given the flux quadrature algorithm. Equations (4.15) can be expressed in a general form for cell i as

$$Q^l_{i+\frac{1}{2}} = Q_i + \frac{1}{4}\left[(1-\kappa)\Delta Q_{i-\frac{1}{2}} + (1+\kappa)\Delta Q_{i+\frac{1}{2}}\right]$$

† The constants a and b should be as small as possible, since the objective is to avoid the formation of new *local* extrema (*i.e.*, unphysical oscillations) in the reconstruction. The minimum values $a = b = 0$, $Q^l_{i+\frac{1}{2}} = Q^r_{i-\frac{1}{2}} = Q_i$ correspond to a first-order reconstruction that is excessively diffusive. Thus, the relevant range is $a \geq 1$ and $b \geq 1$.

$$Q^r_{i-\frac{1}{2}} \;=\; Q_i - \tfrac{1}{4}\Big[(1-\kappa)\Delta Q_{i+\frac{1}{2}} + (1+\kappa)\Delta Q_{i-\frac{1}{2}}\Big] \qquad (4.19)$$

where $\kappa = \frac{1}{3}$. The dependence of $Q^l_{i+\frac{1}{2}}$ and $Q^r_{i-\frac{1}{2}}$ on the adjacent cells is shown in Figs. 4.7 and 4.8. Equation (4.19) is "upwind-biased" since $Q^l_{i+\frac{1}{2}}$ depends on Q_{i-1}, Q_i, and Q_{i+1}, and thus $Q^l_{i+\frac{1}{2}}$ (note *left* face) employs two cells on the *left* of $x_{i+\frac{1}{2}}$ and one cell to the right. Similarly, $Q^r_{i-\frac{1}{2}}$ depends on Q_{i-1}, Q_i, and Q_{i+1}, and thus $Q^r_{i-\frac{1}{2}}$ (note *right* face) employs two cells on the *right* of $x_{i-\frac{1}{2}}$ and one cell to the left. The above formulas can be applied to each cell and thereby define the left and right face values for every face.

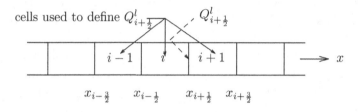

Fig. 4.7. Cells used to define $Q^l_{i+\frac{1}{2}}$ for $\kappa = \frac{1}{3}$

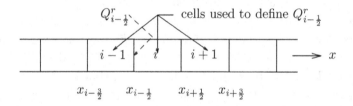

Fig. 4.8. Cells used to define $Q^r_{i-\frac{1}{2}}$ for $\kappa = \frac{1}{3}$

There are three additional interesting values for κ that are summarized in Table 4.2. Equation (4.19) is a second-order reconstruction if $\kappa = -1$ (Exercise 4.1). It is "upwind", *i.e.*, $Q^l_{i+\frac{1}{2}}$ depends only on Q_i and Q_{i-1}, which are to the *left* of $x_{i+\frac{1}{2}}$, as shown in Fig. 4.9, and similarly, $Q^r_{i-\frac{1}{2}}$ depends only on Q_i and Q_{i+1}, which are to the *right* of $x_{i-\frac{1}{2}}$, as illustrated in Fig. 4.10. Equation (4.19) is a second-order upwind-biased reconstruction if $\kappa = 0$ (Exercise 4.4). In this case, the dependence of $Q^l_{i+\frac{1}{2}}$ and $Q^r_{i-\frac{1}{2}}$ on the adjacent cells is the same as shown in Figs. 4.7 and 4.8. Equation (4.19) is also a second-order reconstruction if $\kappa = +1$ (Exercise 4.3). It is a "centered" reconstruction, *i.e.*, $Q^l_{i+\frac{1}{2}}$ depends only on Q_i and Q_{i+1}, which are *centered* on $x_{i+\frac{1}{2}}$, as shown in Fig. 4.11, and similarly $Q^r_{i-\frac{1}{2}}$ depends only on Q_{i-1} and Q_i, which are *centered* on $x_{i-\frac{1}{2}}$, as shown in Fig. 4.12.

Table 4.2. *Reconstruction*

κ	Order	Definition
1	2nd	Centered
$\frac{1}{3}$	3rd	Upwind-biased
0	2nd	Upwind-biased
-1	2nd	Upwind

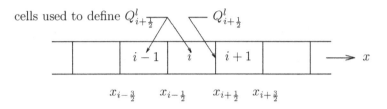

Fig. 4.9. Cells used to define $Q^l_{i+\frac{1}{2}}$ for $\kappa = -1$

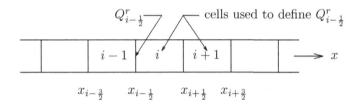

Fig. 4.10. Cells used to define $Q^r_{i-\frac{1}{2}}$ for $\kappa = -1$

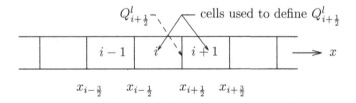

Fig. 4.11. Cells used to define $Q^l_{i+\frac{1}{2}}$ for $\kappa = +1$

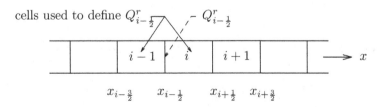

Fig. 4.12. Cells used to define $Q^r_{i-\frac{1}{2}}$ for $\kappa = +1$

Anderson *et al.* (1986) modified (4.19) by replacing $\Delta Q_{i\pm\frac{1}{2}}$ with $\widehat{\Delta Q}_{i\pm\frac{1}{2}}$:

$$
\begin{aligned}
Q^l_{i+\frac{1}{2}} &= Q_i + \tfrac{1}{4}\left[(1-\kappa)\widehat{\Delta Q}_{i-\frac{1}{2}} + (1+\kappa)\widehat{\Delta Q}_{i+\frac{1}{2}}\right] \\
Q^r_{i-\frac{1}{2}} &= Q_i - \tfrac{1}{4}\left[(1-\kappa)\widehat{\Delta Q}_{i+\frac{1}{2}} + (1+\kappa)\widehat{\Delta Q}_{i-\frac{1}{2}}\right] \quad (4.20)
\end{aligned}
$$

where $\widehat{\Delta Q}_{i\pm\frac{1}{2}} = \Delta Q_{i\pm\frac{1}{2}}$ except when this would introduce new extrema in $Q^l_{i+\frac{1}{2}}$ or $Q^r_{i-\frac{1}{2}}$ (*i.e.*, a new minimum or maximum) relative to Q_{i-1}, Q_i and Q_{i+1}. This implies four constraints:

$$Q^l_{i+\frac{1}{2}} \leq \max(Q_{i-1}, Q_i, Q_{i+1}) \qquad (4.21)$$

$$Q^l_{i+\frac{1}{2}} \geq \min(Q_{i-1}, Q_i, Q_{i+1}) \qquad (4.22)$$

$$Q^r_{i-\frac{1}{2}} \leq \max(Q_{i-1}, Q_i, Q_{i+1}) \qquad (4.23)$$

$$Q^r_{i-\frac{1}{2}} \geq \min(Q_{i-1}, Q_i, Q_{i+1}) \qquad (4.24)$$

This approach is denoted *Modified Upwind Scheme for Conservation Laws (MUSCL)*.[†] There are four cases to consider for the values of κ in Table 4.2.

4.4.1 Case 1: $\Delta Q_{i+\frac{1}{2}} \geq 0$, $\Delta Q_{i-\frac{1}{2}} \geq 0$

The relative values of Q_{i-1}, Q_i, and Q_{i+1} are illustrated in Fig. 4.13. The first constraint (4.21) yields

$$Q_i + \tfrac{1}{4}\left[(1-\kappa)\widehat{\Delta Q}_{i-\frac{1}{2}} + (1+\kappa)\widehat{\Delta Q}_{i+\frac{1}{2}}\right] \leq Q_{i+1}$$

using (4.20). Assume that $\widehat{\Delta Q}_{i\pm\frac{1}{2}} = \Delta Q_{i\pm\frac{1}{2}}$. Then the above condition becomes

$$\Delta Q_{i-\frac{1}{2}} \leq b\,\Delta Q_{i+\frac{1}{2}}$$

where b is a positive constant defined for $\kappa < 1$ by[‡]

$$b = \frac{3-\kappa}{1-\kappa} \qquad (4.25)$$

Thus, provided $\Delta Q_{i-\frac{1}{2}} \leq b\,\Delta Q_{i+\frac{1}{2}}$, we can assign $\widehat{\Delta Q}_{i\pm\frac{1}{2}} = \Delta Q_{i\pm\frac{1}{2}}$. When $\Delta Q_{i-\frac{1}{2}} > b\,\Delta Q_{i+\frac{1}{2}}$, we write

$$\Delta Q_{i-\frac{1}{2}} = b\,\Delta Q_{i+\frac{1}{2}} + \delta$$

---

† Earlier, Van Leer (1977a,1977b,1979) developed a combined reconstruction-evolution method entitled Monotone Upwind Schemes for Scalar Conservation Laws denoted by the same acronym (MUSCL). Hirsch (1988) uses the term Monotone Upstream-Centred Schemes for Conservation Laws.

‡ If $\kappa = 1$, then (4.21) is satisfied automatically.

where $\delta \geq 0$. Substituting

$$Q^l_{i+\frac{1}{2}} = Q_{i+1} + \frac{(1-\kappa)}{4}\delta$$

and δ must therefore be zero. Therefore, the first constraint (4.21) is satisfied provided

$$
\begin{aligned}
\widehat{\Delta Q}_{i-\frac{1}{2}} &= \begin{cases} \Delta Q_{i-\frac{1}{2}} & \text{if } \Delta Q_{i-\frac{1}{2}} \leq b\,\Delta Q_{i+\frac{1}{2}} \\ b\,\Delta Q_{i+\frac{1}{2}} & \text{if } \Delta Q_{i-\frac{1}{2}} > b\,\Delta Q_{i+\frac{1}{2}} \end{cases} \\
\widehat{\Delta Q}_{i+\frac{1}{2}} &= \Delta Q_{i+\frac{1}{2}}
\end{aligned}
\tag{4.26}
$$

If $\kappa = 1$, then $\widehat{\Delta Q}_{i\pm\frac{1}{2}} = \Delta Q_{i\pm\frac{1}{2}}$.

The second constraint (4.22) yields

$$Q_i + \frac{1}{4}\left[(1-\kappa)\widehat{\Delta Q}_{i-\frac{1}{2}} + (1+\kappa)\widehat{\Delta Q}_{i+\frac{1}{2}}\right] \geq Q_{i-1}$$

Assuming $\widehat{\Delta Q}_{i\pm\frac{1}{2}} = \Delta Q_{i\pm\frac{1}{2}}$, this constraint becomes

$$(5-\kappa)\Delta Q_{i-\frac{1}{2}} + (1+\kappa)\Delta Q_{i+\frac{1}{2}} \geq 0$$

Since $\Delta Q_{i\pm\frac{1}{2}} \geq 0$, this condition is satisfied.

The third constraint (4.23) yields

$$Q_i - \frac{1}{4}\left[(1-\kappa)\widehat{\Delta Q}_{i+\frac{1}{2}} + (1+\kappa)\widehat{\Delta Q}_{i-\frac{1}{2}}\right] \leq Q_{i+1}$$

Assuming $\widehat{\Delta Q}_{i\pm\frac{1}{2}} = \Delta Q_{i\pm\frac{1}{2}}$, this constraint becomes

$$(5-\kappa)\Delta Q_{i+\frac{1}{2}} + (1+\kappa)\Delta Q_{i-\frac{1}{2}} \geq 0$$

which is satisfied.

The fourth constraint (4.24) yields

$$Q_i - \frac{1}{4}\left[(1-\kappa)\widehat{\Delta Q}_{i+\frac{1}{2}} + (1+\kappa)\widehat{\Delta Q}_{i-\frac{1}{2}}\right] \geq Q_{i-1}$$

Assuming $\widehat{\Delta Q}_{i\pm\frac{1}{2}} = \Delta Q_{i\pm\frac{1}{2}}$, this constraint becomes, for $\kappa < 1$,

$$\Delta Q_{i+\frac{1}{2}} \leq b\,\Delta Q_{i-\frac{1}{2}}$$

By a similar argument as employed for the first constraint, we obtain

$$
\begin{aligned}
\widehat{\Delta Q}_{i-\frac{1}{2}} &= \Delta Q_{i-\frac{1}{2}} \\
\widehat{\Delta Q}_{i+\frac{1}{2}} &= \begin{cases} \Delta Q_{i+\frac{1}{2}} & \text{if } \Delta Q_{i+\frac{1}{2}} \leq b\,\Delta Q_{i-\frac{1}{2}} \\ b\,\Delta Q_{i-\frac{1}{2}} & \text{if } \Delta Q_{i+\frac{1}{2}} > b\,\Delta Q_{i-\frac{1}{2}} \end{cases}
\end{aligned}
\tag{4.27}
$$

If $\kappa = 1$, then $\widehat{\Delta Q}_{i\pm\frac{1}{2}} = \Delta Q_{i\pm\frac{1}{2}}$.

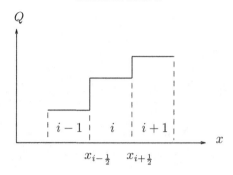

Fig. 4.13. $\Delta Q_{i+\frac{1}{2}} \geq 0,\ \Delta Q_{i-\frac{1}{2}} \geq 0$

Combining (4.26) and (4.27) yields, for $\kappa < 1$,

$$\widehat{\Delta Q}_{i-\frac{1}{2}} = \begin{cases} \Delta Q_{i-\frac{1}{2}} & \text{if } \Delta Q_{i-\frac{1}{2}} \leq b\,\Delta Q_{i+\frac{1}{2}} \\ b\,\Delta Q_{i+\frac{1}{2}} & \text{if } \Delta Q_{i-\frac{1}{2}} > b\,\Delta Q_{i+\frac{1}{2}} \end{cases}$$

$$\widehat{\Delta Q}_{i+\frac{1}{2}} = \begin{cases} \Delta Q_{i+\frac{1}{2}} & \text{if } \Delta Q_{i+\frac{1}{2}} \leq b\,\Delta Q_{i-\frac{1}{2}} \\ b\,\Delta Q_{i-\frac{1}{2}} & \text{if } \Delta Q_{i+\frac{1}{2}} > b\,\Delta Q_{i-\frac{1}{2}} \end{cases} \tag{4.28}$$

and $\widehat{\Delta Q}_{i\pm\frac{1}{2}} = \Delta Q_{i\pm\frac{1}{2}}$ for $\kappa = 1$. This is illustrated in Figs. 4.14 to 4.16 for $\kappa < 1$.

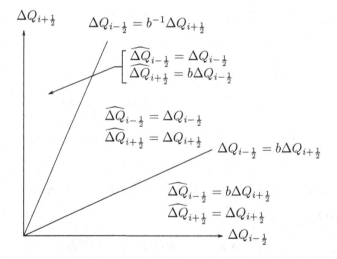

Fig. 4.14. $\widehat{\Delta Q}_{i+\frac{1}{2}}$ and $\widehat{\Delta Q}_{i-\frac{1}{2}}$ for Case 1

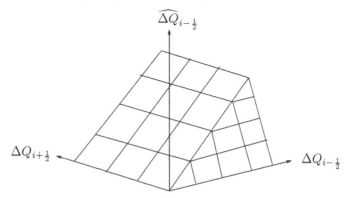

Fig. 4.15. $\widehat{\Delta Q}_{i-\frac{1}{2}}$ for Case 1

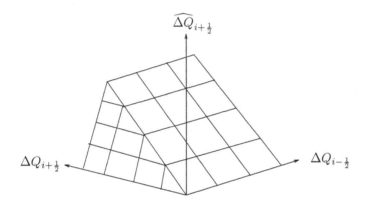

Fig. 4.16. $\widehat{\Delta Q}_{i+\frac{1}{2}}$ for Case 1

4.4.2 Case 2: $\Delta Q_{i+\frac{1}{2}} \geq 0,\ \Delta Q_{i-\frac{1}{2}} \leq 0$

The relative values are illustrated in Fig. 4.17, which is shown for $\Delta Q_{i+\frac{1}{2}} > -\Delta Q_{i-\frac{1}{2}}$. We consider the four values of κ in Table 4.2.

For $\kappa = -1$, Equation (4.20) is

$$
\begin{aligned}
Q^l_{i+\frac{1}{2}} &= Q_i + \tfrac{1}{2}\widehat{\Delta Q}_{i-\frac{1}{2}} \\
Q^r_{i+\frac{1}{2}} &= Q_i - \tfrac{1}{2}\widehat{\Delta Q}_{i+\frac{1}{2}}
\end{aligned}
$$

Assume that $\widehat{\Delta Q}_{i\pm\frac{1}{2}} = \Delta Q_{i\pm\frac{1}{2}}$. Equations (4.22) and (4.24) imply

$$
\begin{aligned}
\Delta Q_{i-\frac{1}{2}} &\geq 0 \\
\Delta Q_{i+\frac{1}{2}} &\leq 0
\end{aligned}
$$

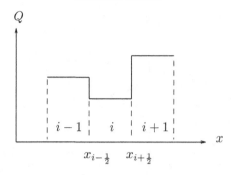

Fig. 4.17. $\Delta Q_{i+\frac{1}{2}} \geq 0$, $\Delta Q_{i-\frac{1}{2}} \leq 0$

which is opposite of the conditions for Case 2. Thus, we take

$$\widehat{\Delta Q}_{i-\frac{1}{2}} = 0$$
$$\widehat{\Delta Q}_{i+\frac{1}{2}} = 0$$

For $\kappa = 0$, Equation (4.20) is

$$Q^l_{i+\frac{1}{2}} = Q_i + \tfrac{1}{4}\left(\widehat{\Delta Q}_{i+\frac{1}{2}} + \widehat{\Delta Q}_{i-\frac{1}{2}}\right)$$
$$Q^r_{i+\frac{1}{2}} = Q_i - \tfrac{1}{4}\left(\widehat{\Delta Q}_{i+\frac{1}{2}} + \widehat{\Delta Q}_{i-\frac{1}{2}}\right)$$

Assume that $\widehat{\Delta Q}_{i\pm\frac{1}{2}} = \Delta Q_{i\pm\frac{1}{2}}$. Equation (4.22) implies

$$\Delta Q_{i+\frac{1}{2}} + \Delta Q_{i-\frac{1}{2}} \geq 0$$

while Equation (4.24) implies

$$\Delta Q_{i+\frac{1}{2}} + \Delta Q_{i-\frac{1}{2}} \leq 0$$

which are in general contradictory. Thus, we take

$$\widehat{\Delta Q}_{i-\frac{1}{2}} = 0$$
$$\widehat{\Delta Q}_{i+\frac{1}{2}} = 0$$

For $\kappa = \frac{1}{3}$, Equation (4.20) is

$$Q^l_{i+\frac{1}{2}} = Q_i + \tfrac{1}{6}\left(\widehat{\Delta Q}_{i-\frac{1}{2}} + 2\widehat{\Delta Q}_{i+\frac{1}{2}}\right)$$
$$Q^r_{i+\frac{1}{2}} = Q_i - \tfrac{1}{6}\left(\widehat{\Delta Q}_{i+\frac{1}{2}} + 2\widehat{\Delta Q}_{i-\frac{1}{2}}\right)$$

Assume that $\widehat{\Delta Q}_{i\pm\frac{1}{2}} = \Delta Q_{i\pm\frac{1}{2}}$ except where a limiter must be employed.

The following definitions satisfy Equations (4.21) to (4.24):

$$\widehat{\Delta Q}_{i-\frac{1}{2}} = \begin{cases} \Delta Q_{i-\frac{1}{2}} & \text{if } \Delta Q_{i-\frac{1}{2}} \geq -2\,\Delta Q_{i+\frac{1}{2}} \\ -2\,\Delta Q_{i+\frac{1}{2}} & \text{if } \Delta Q_{i-\frac{1}{2}} < -2\,\Delta Q_{i+\frac{1}{2}} \end{cases}$$

$$\widehat{\Delta Q}_{i+\frac{1}{2}} = \begin{cases} \Delta Q_{i+\frac{1}{2}} & \text{if } \Delta Q_{i+\frac{1}{2}} \leq -2\,\Delta Q_{i-\frac{1}{2}} \\ -2\,\Delta Q_{i-\frac{1}{2}} & \text{if } \Delta Q_{i+\frac{1}{2}} > -2\,\Delta Q_{i-\frac{1}{2}} \end{cases}$$

For $\kappa = 1$, Equation (4.20) is

$$Q^{l}_{i+\frac{1}{2}} = Q_i + \tfrac{1}{2}\widehat{\Delta Q}_{i+\frac{1}{2}}$$

$$Q^{r}_{i+\frac{1}{2}} = Q_i - \tfrac{1}{2}\widehat{\Delta Q}_{i-\frac{1}{2}}$$

Assume that $\widehat{\Delta Q}_{i\pm\frac{1}{2}} = \Delta Q_{i\pm\frac{1}{2}}$. Then

$$Q^{l}_{i+\frac{1}{2}} = Q_i + \tfrac{1}{2}\Delta Q_{i+\frac{1}{2}}$$

$$Q^{r}_{i+\frac{1}{2}} = Q_i - \tfrac{1}{2}\Delta Q_{i-\frac{1}{2}}$$

which satisfy the conditions in Equations (4.21) to (4.24).

4.4.3 Case 3: $\Delta Q_{i+\frac{1}{2}} \leq 0$, $\Delta Q_{i-\frac{1}{2}} \leq 0$

The relative values are illustrated in Fig. 4.18. The analysis is similar to Case 1. The result for $\kappa < 1$ is

$$\widehat{\Delta Q}_{i-\frac{1}{2}} = \begin{cases} \Delta Q_{i-\frac{1}{2}} & \text{if } \Delta Q_{i-\frac{1}{2}} \geq b\,\Delta Q_{i+\frac{1}{2}} \\ b\,\Delta Q_{i+\frac{1}{2}} & \text{if } \Delta Q_{i-\frac{1}{2}} < b\,\Delta Q_{i+\frac{1}{2}} \end{cases}$$

$$\widehat{\Delta Q}_{i+\frac{1}{2}} = \begin{cases} \Delta Q_{i+\frac{1}{2}} & \text{if } \Delta Q_{i+\frac{1}{2}} \geq b\,\Delta Q_{i-\frac{1}{2}} \\ b\,\Delta Q_{i-\frac{1}{2}} & \text{if } \Delta Q_{i+\frac{1}{2}} < b\,\Delta Q_{i-\frac{1}{2}} \end{cases} \tag{4.29}$$

and $\widehat{\Delta Q}_{i\pm\frac{1}{2}} = \Delta Q_{i\pm\frac{1}{2}}$ for $\kappa = 1$.

4.4.4 Case 4: $\Delta Q_{i+\frac{1}{2}} \leq 0$, $\Delta Q_{i-\frac{1}{2}} \geq 0$

The relative values are illustrated in Fig. 4.19. The analysis is similar to Case 2. For $\kappa = -1$ and 0, the result is

$$\widehat{\Delta Q}_{i-\frac{1}{2}} = 0$$

$$\widehat{\Delta Q}_{i+\frac{1}{2}} = 0$$

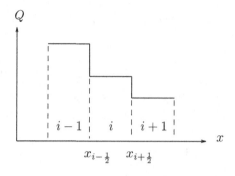

Fig. 4.18. $\Delta Q_{i+\frac{1}{2}} \leq 0$, $\Delta Q_{i-\frac{1}{2}} \leq 0$

For $\kappa = \frac{1}{3}$, the result is

$$\widehat{\Delta Q}_{i-\frac{1}{2}} = \begin{cases} \Delta Q_{i-\frac{1}{2}} & \text{if } \Delta Q_{i-\frac{1}{2}} \leq -2\,\Delta Q_{i+\frac{1}{2}} \\ -2\,\Delta Q_{i+\frac{1}{2}} & \text{if } \Delta Q_{i-\frac{1}{2}} > -2\,\Delta Q_{i+\frac{1}{2}} \end{cases}$$

$$\widehat{\Delta Q}_{i+\frac{1}{2}} = \begin{cases} \Delta Q_{i+\frac{1}{2}} & \text{if } \Delta Q_{i+\frac{1}{2}} \geq -2\,\Delta Q_{i-\frac{1}{2}} \\ -2\,\Delta Q_{i-\frac{1}{2}} & \text{if } \Delta Q_{i+\frac{1}{2}} < -2\,\Delta Q_{i-\frac{1}{2}} \end{cases}$$

For $\kappa = 1$, the result is

$$\widehat{\Delta Q}_{i+\frac{1}{2}} = \Delta Q_{i+\frac{1}{2}}$$

$$\widehat{\Delta Q}_{i-\frac{1}{2}} = \Delta Q_{i-\frac{1}{2}}$$

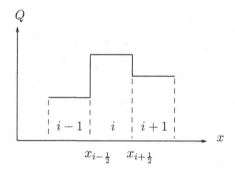

Fig. 4.19. $\Delta Q_{i+\frac{1}{2}} \leq 0$, $\Delta Q_{i-\frac{1}{2}} \geq 0$

4.4.5 Summary

The MUSCL scheme is

$$Q^l_{i+\frac{1}{2}} = Q_i + \frac{1}{4}\left[(1-\kappa)\widehat{\Delta Q}_{i-\frac{1}{2}} + (1+\kappa)\widehat{\Delta Q}_{i+\frac{1}{2}}\right]$$

$$Q^r_{i-\frac{1}{2}} = Q_i - \frac{1}{4}\left[(1-\kappa)\widehat{\Delta Q}_{i+\frac{1}{2}} + (1+\kappa)\widehat{\Delta Q}_{i-\frac{1}{2}}\right]$$

For $\kappa = 1$,

$$\widehat{\Delta Q}_{i-\frac{1}{2}} = \Delta Q_{i-\frac{1}{2}}$$
$$\widehat{\Delta Q}_{i+\frac{1}{2}} = \Delta Q_{i+\frac{1}{2}}$$

$$(4.30)$$

For $\kappa = \frac{1}{3}$ with $\Delta Q_{i+\frac{1}{2}} \geq 0$ and $\Delta Q_{i-\frac{1}{2}} \geq 0$,

$$\widehat{\Delta Q}_{i-\frac{1}{2}} = \begin{cases} \Delta Q_{i-\frac{1}{2}} & \text{if } \Delta Q_{i-\frac{1}{2}} \leq b\,\Delta Q_{i+\frac{1}{2}} \\ b\,\Delta Q_{i+\frac{1}{2}} & \text{if } \Delta Q_{i-\frac{1}{2}} > b\,\Delta Q_{i+\frac{1}{2}} \end{cases}$$

$$\widehat{\Delta Q}_{i+\frac{1}{2}} = \begin{cases} \Delta Q_{i+\frac{1}{2}} & \text{if } \Delta Q_{i+\frac{1}{2}} \leq b\,\Delta Q_{i-\frac{1}{2}} \\ b\,\Delta Q_{i-\frac{1}{2}} & \text{if } \Delta Q_{i+\frac{1}{2}} > b\,\Delta Q_{i-\frac{1}{2}} \end{cases}$$

For $\kappa = \frac{1}{3}$ with $\Delta Q_{i+\frac{1}{2}} \geq 0$ and $\Delta Q_{i-\frac{1}{2}} \leq 0$,

$$\widehat{\Delta Q}_{i-\frac{1}{2}} = \begin{cases} \Delta Q_{i-\frac{1}{2}} & \text{if } \Delta Q_{i-\frac{1}{2}} \geq -2\,\Delta Q_{i+\frac{1}{2}} \\ -2\,\Delta Q_{i+\frac{1}{2}} & \text{if } \Delta Q_{i-\frac{1}{2}} < -2\,\Delta Q_{i+\frac{1}{2}} \end{cases}$$

$$\widehat{\Delta Q}_{i+\frac{1}{2}} = \begin{cases} \Delta Q_{i+\frac{1}{2}} & \text{if } \Delta Q_{i+\frac{1}{2}} \leq -2\,\Delta Q_{i-\frac{1}{2}} \\ -2\,\Delta Q_{i-\frac{1}{2}} & \text{if } \Delta Q_{i+\frac{1}{2}} > -2\,\Delta Q_{i-\frac{1}{2}} \end{cases}$$

For $\kappa = \frac{1}{3}$ with $\Delta Q_{i+\frac{1}{2}} \leq 0$ and $\Delta Q_{i-\frac{1}{2}} \leq 0$,

$$\widehat{\Delta Q}_{i-\frac{1}{2}} = \begin{cases} \Delta Q_{i-\frac{1}{2}} & \text{if } \Delta Q_{i-\frac{1}{2}} \geq b\,\Delta Q_{i+\frac{1}{2}} \\ b\,\Delta Q_{i+\frac{1}{2}} & \text{if } \Delta Q_{i-\frac{1}{2}} < b\,\Delta Q_{i+\frac{1}{2}} \end{cases}$$

$$\widehat{\Delta Q}_{i+\frac{1}{2}} = \begin{cases} \Delta Q_{i+\frac{1}{2}} & \text{if } \Delta Q_{i+\frac{1}{2}} \geq b\,\Delta Q_{i-\frac{1}{2}} \\ b\,\Delta Q_{i-\frac{1}{2}} & \text{if } \Delta Q_{i+\frac{1}{2}} < b\,\Delta Q_{i-\frac{1}{2}} \end{cases}$$

For $\kappa = \frac{1}{3}$ with $\Delta Q_{i+\frac{1}{2}} \leq 0$ and $\Delta Q_{i-\frac{1}{2}} \geq 0$,

$$\widehat{\Delta Q}_{i-\frac{1}{2}} = \begin{cases} \Delta Q_{i-\frac{1}{2}} & \text{if } \Delta Q_{i-\frac{1}{2}} \leq -2\,\Delta Q_{i+\frac{1}{2}} \\ -2\,\Delta Q_{i+\frac{1}{2}} & \text{if } \Delta Q_{i-\frac{1}{2}} > -2\,\Delta Q_{i+\frac{1}{2}} \end{cases}$$

$$\widehat{\Delta Q}_{i+\frac{1}{2}} = \begin{cases} \Delta Q_{i+\frac{1}{2}} & \text{if } \Delta Q_{i+\frac{1}{2}} \geq -2\,\Delta Q_{i-\frac{1}{2}} \\ -2\,\Delta Q_{i-\frac{1}{2}} & \text{if } \Delta Q_{i+\frac{1}{2}} < -2\,\Delta Q_{i-\frac{1}{2}} \end{cases}$$

For $\kappa = 0$ and -1 (Exercise 4.7),

$$\widehat{\Delta Q}_{i-\frac{1}{2}} = \text{minmod}\,(\Delta Q_{i-\frac{1}{2}}, b\Delta Q_{i+\frac{1}{2}})$$
$$\widehat{\Delta Q}_{i+\frac{1}{2}} = \text{minmod}\,(b\Delta Q_{i-\frac{1}{2}}, \Delta Q_{i+\frac{1}{2}})$$

where

$$\mathrm{minmod}\,(x,y) = \begin{cases} x & \text{if } |x| \le |y| \text{ where } x \text{ and } y \text{ have the same sign} \\ y & \text{if } |x| > |y| \text{ where } x \text{ and } y \text{ have the same sign} \\ 0 & \text{where } x \text{ and } y \text{ have opposite signs} \end{cases}$$

4.4.6 Results

The effect of the limiter on the reconstruction for the sin wave with ten cells is shown in Fig. 4.20 using (4.20) with $\kappa = \frac{1}{3}$. The reconstructed values at the left and right sides of each face are identical to Fig. 4.5, and thus the limiter has no effect in this case. Similar results[1] are obtained for $\kappa = -1, 0$, and 1.

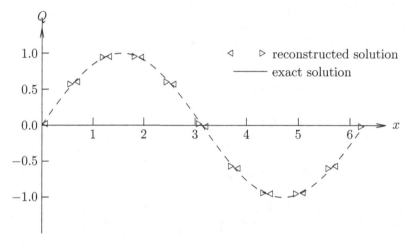

Fig. 4.20. Reconstruction of sine using ten cells and $\kappa = \frac{1}{3}$ with a limiter

The effect of the limiter on the reconstruction for the discontinuity (4.18) with ten cells is shown in Fig. 4.21 using (4.20) with $\kappa = \frac{1}{3}$. The limiter yields an accurate reconstruction of the left and right states within each cell including the discontinuity at $x = \pi$. Identical results are obtained with $\kappa = -1$ and 0. However, the left and right states at the discontinuity are incorrectly reconstructed for $\kappa = 1$, as shown in Fig. 4.22.

4.5 Essentially Non-Oscillatory Methods

In Section 4.2, the reconstruction polynomial $Q_i(x)$ within cell i, given by Equation (4.14), uses the cell average values Q_{i-1}, Q_i, and Q_{i+1} as illustrated

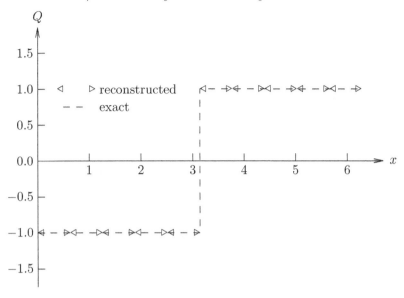

Fig. 4.21. Reconstruction using ten cells and $\kappa = \frac{1}{3}$ with a limiter

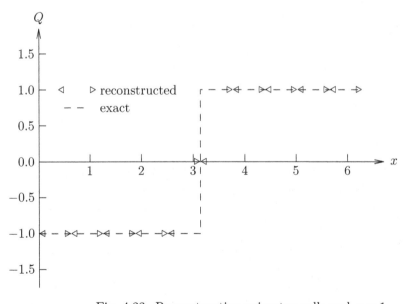

Fig. 4.22. Reconstruction using ten cells and $\kappa = 1$

in Fig. 4.3. This stencil of cells is symmetric about cell i and fixed. For smooth functions, this leads to an accurate reconstruction within the cell i (*e.g.*, Fig. 4.5). However, for discontinuous functions, the use of a fixed symmetric stencil of cells in the vicinity of the discontinuity yields unphysical extrema (*e.g.*, Fig. 4.6). The explanation is seen in Fig. 4.23. Reconstruction of the Q in cell number 5 should utilize only information from cell number 5 and cells to its left. Similarly, reconstruction of Q in cell number 6 should utilize only information from cell number 6 and cells to its right. Use of a symmetric stencil for reconstruction of Q within cells number 5 and 6 leads to unphysical extrema as indicated in Fig. 4.6. The MUSCL scheme described in the previous section eliminates the unphysical extrema as indicated in Figs 4.21 and 4.22 while retaining a symmetric stencil.

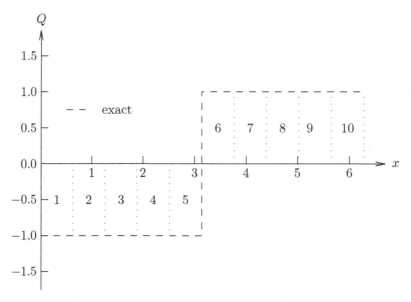

Fig. 4.23. Discontinuous function with ten cells

Harten *et al.* (1987a) developed a reconstruction method that permits an asymmetric stencil of cells. We develop the method for the case of a second-order accurate reconstruction of $Q^l_{i+\frac{1}{2}}$ and $Q^r_{i-\frac{1}{2}}$ in cell i. The extension to higher order is described in Harten *et al.* (1987b). The reconstruction polynomial $Q_i(x)$ is required to satisfy three conditions. First,

$$Q(x) = \mathcal{Q}(x) + \mathcal{O}(\Delta x^3) \tag{4.31}$$

in regions where $\mathcal{Q}(x)$ is smooth. Second, $Q_i(x)$ reconstructs the cell average exactly,

$$Q_i = \int_{x_{i-\frac{1}{2}}}^{x_{i+\frac{1}{2}}} Q_i(x)\, dx \tag{4.32}$$

Third, $Q_i(x)$ satisfies the Essentially Non-Oscillatory property

$$TV(Q_i(x)) \leq TV(\mathcal{Q}) + \mathcal{O}(\Delta x^3) \tag{4.33}$$

where TV is the Total Variation of a function defined as (Hirsch, 1988)

$$TV(Q) = \int_{x_{i-\frac{1}{2}}}^{x_{i+\frac{1}{2}}} \left| \frac{dQ}{dx} \right| dx \tag{4.34}$$

and constitutes a measure of the oscillatory behavior of $Q_i(x)$ within cell i (see also Section 8.2).

Following Section 4.2, we define the primitive function

$$I(x) = \int_{x_{i-a-\frac{1}{2}}}^{x} Q \, dx \quad \text{for} \quad x_{i-a-\frac{1}{2}} \leq x \leq x_{i-a+\frac{5}{2}} \tag{4.35}$$

Note that $I(x)$ is defined for x within the limits of the three cells $i-a$, $i-a+1$, and $i-a+2$, and Equation (4.2) corresponds to $a = 1$ in Equation (4.35). Then, similar to Equation (4.3),

$$
\begin{aligned}
I(x_{i-a-\frac{1}{2}}) &= 0 \\
I(x_{i-a+\frac{1}{2}}) &= \Delta x \, Q_{i-a} \\
I(x_{i-a+\frac{3}{2}}) &= \Delta x \, (Q_{i-a} + Q_{i-a+1}) \\
I(x_{i-a+\frac{5}{2}}) &= \Delta x \, (Q_{i-a} + Q_{i-a+1} + Q_{i-a+2})
\end{aligned}
\tag{4.36}
$$

There is a unique third-order polynomial $P(x)$ that interpolates $I(x)$ at the four points $x_{i-a-\frac{1}{2}}$, $x_{i-a+\frac{1}{2}}$, $x_{i-a+\frac{3}{2}}$, and $x_{i-a+\frac{5}{2}}$. The polynomial may be obtained using Newton's formula (Mathews and Fink, 1998)

$$
\begin{aligned}
P(x) = \; & a_0 + a_1(x - x_{i-a-\frac{1}{2}}) + a_2(x - x_{i-a-\frac{1}{2}})(x - x_{i-a+\frac{1}{2}}) + \\
& a_3(x - x_{i-a-\frac{1}{2}})(x - x_{i-a+\frac{1}{2}})(x - x_{i-a+\frac{3}{2}})
\end{aligned}
$$

The coefficients are defined by

$$
\begin{aligned}
a_0 &= I[x_{i-a-\frac{1}{2}}] \\
a_1 &= I[x_{i-a-\frac{1}{2}}, x_{i-a+\frac{1}{2}}] \\
a_2 &= I[x_{i-a-\frac{1}{2}}, x_{i-a+\frac{1}{2}}, x_{i-a+\frac{3}{2}}] \\
a_3 &= I[x_{i-a-\frac{1}{2}}, x_{i-a+\frac{1}{2}}, x_{i-a+\frac{3}{2}}, x_{i-a+\frac{5}{2}}]
\end{aligned}
\tag{4.37}
$$

where

$$
\begin{aligned}
I[x_{i-a-\frac{1}{2}}] &= I(x_{i-a-\frac{1}{2}}) \\
I[x_{i-a-\frac{1}{2}}, x_{i-a+\frac{1}{2}}] &= \frac{I[x_{i-a+\frac{1}{2}}] - I[x_{i-a-\frac{1}{2}}]}{x_{i-a+\frac{1}{2}} - x_{i-a-\frac{1}{2}}} \\
I[x_{i-a-\frac{1}{2}}, x_{i-a+\frac{1}{2}}, x_{i-a+\frac{3}{2}}] &= \frac{I[x_{i-a+\frac{1}{2}}, x_{i-a+\frac{3}{2}}] - I[x_{i-a-\frac{1}{2}}, x_{i-a+\frac{1}{2}}]}{x_{i-a+\frac{3}{2}} - x_{i-a-\frac{1}{2}}}
\end{aligned}
$$

$$I[x_{i-a-\frac{1}{2}}, x_{i-a+\frac{1}{2}}, x_{i-a+\frac{3}{2}}, x_{i-a+\frac{5}{2}}] = \left(x_{i-a+\frac{5}{2}} - x_{i-a-\frac{1}{2}}\right)^{-1}$$
$$\left(I[x_{i-a+\frac{1}{2}}, x_{i-a+\frac{3}{2}}, x_{i-a+\frac{5}{2}}] -\right.$$
$$\left. I[x_{i-a-\frac{1}{2}}, x_{i-a+\frac{1}{2}}, x_{i-a+\frac{3}{2}}]\right)$$

$$(4.38)$$

Now

$$I(x) = P(x) + E(x) \quad \text{for} \quad x_{i-a-\frac{1}{2}} \le x \le x_{i-a+\frac{5}{2}} \tag{4.39}$$

where $E(x)$ is the error defined by

$$E(x) = \frac{(x - x_{i-a-\frac{1}{2}})(x - x_{i-a+\frac{1}{2}})(x - x_{i-a+\frac{3}{2}})(x - x_{i-a+\frac{5}{2}})}{4!} \frac{d^4 I}{dx^4}\bigg|_{x=\hat{x}} \tag{4.40}$$

where $\hat{x}$ depends on x and $x_{i-a-\frac{1}{2}} \le \hat{x} \le x_{i-a+\frac{5}{2}}$.

The reconstruction function for $\mathcal{Q}$ in cell i is denoted as $Q_i(x)$ and is defined by

$$Q_i(x) = \frac{dP}{dx} \quad \text{for} \quad x_{i-\frac{1}{2}} \le x \le x_{i+\frac{1}{2}} \tag{4.41}$$

Thus

$$Q_i(x) = a_1 + a_2 \left[(x - x_{i-a+\frac{1}{2}}) + (x - x_{i-a-\frac{1}{2}})\right] +$$
$$a_3 \left[(x - x_{i-a+\frac{1}{2}})(x + x_{i-a+\frac{3}{2}}) + (x - x_{i-a-\frac{1}{2}})(x - x_{i-a+\frac{3}{2}}) +\right.$$
$$\left. (x - x_{i-a-\frac{1}{2}})(x - x_{i-a+\frac{1}{2}})\right] \tag{4.42}$$

Using (4.36), (4.37), and (4.38),

$$I[x_{i-a-\frac{1}{2}}, x_{i-a+\frac{1}{2}}] = Q_{i-a}$$
$$I[x_{i-a+\frac{1}{2}}, x_{i-a+\frac{3}{2}}] = Q_{i-a+1}$$
$$I[x_{i-a+\frac{3}{2}}, x_{i-a+\frac{5}{2}}] = Q_{i-a+2}$$
$$I[x_{i-a-\frac{1}{2}}, x_{i-a+\frac{1}{2}}, x_{i-a+\frac{3}{2}}] = (2\Delta x)^{-1} (Q_{i-a+1} - Q_{i-a})$$
$$I[x_{i-a+\frac{1}{2}}, x_{i-a+\frac{3}{2}}, x_{i-a+\frac{5}{2}}] = (2\Delta x)^{-1} (Q_{i-a+2} - Q_{i-a+1}) \tag{4.43}$$

Using (4.11), the coefficients are

$$a_1 = Q_{i-a}$$
$$a_2 = (2\Delta x)^{-1} \Delta Q_{i-a+\frac{1}{2}}$$
$$a_3 = \left[6(\Delta x)^2\right]^{-1} \left(\Delta Q_{i-a+\frac{3}{2}} - \Delta Q_{i-a+\frac{1}{2}}\right) \tag{4.44}$$

Using[2] (4.13),

$$Q_i(x) = Q_i - \frac{(\Delta Q_{i-a+\frac{3}{2}} - \Delta Q_{i-a+\frac{1}{2}})}{24}$$

$$+ \frac{1}{\Delta x} \left[\Delta Q_{i-a+\frac{1}{2}} + \left(a - \tfrac{1}{2}\right)\left(\Delta Q_{i-a+\frac{3}{2}} - \Delta Q_{i-a+\frac{1}{2}}\right) \right] \xi$$

$$+ \frac{(\Delta Q_{i-a+\frac{3}{2}} - \Delta Q_{i-a+\frac{1}{2}})}{2(\Delta x)^2} \xi^2 \qquad (4.45)$$

where $\xi = x - x_i$.

From (4.35), (4.39), and (4.40),

$$Q_i(x) = \frac{dI}{dx} - \frac{dE}{dx}$$

$$= \mathcal{Q}(x) - \frac{dE}{dx}$$

From (4.40), it is evident that $dE/dx = \mathcal{O}(\Delta x^3)$. Therefore, $Q_i(x)$ is a third-order accurate reconstruction of $\mathcal{Q}$. It is also straightforward to verify (4.33) since

$$\int_{x_{i-\frac{1}{2}}}^{x_{i+\frac{1}{2}}} \left| \frac{dQ_i}{dx} \right| dx = \int_{x_{i-\frac{1}{2}}}^{x_{i+\frac{1}{2}}} \left| \frac{d\mathcal{Q}}{dx} - \frac{d^2E}{dx^2} \right| dx$$

$$\leq \int_{x_{i-\frac{1}{2}}}^{x_{i+\frac{1}{2}}} \left| \frac{d\mathcal{Q}}{dx} \right| dx + \int_{x_{i-\frac{1}{2}}}^{x_{i+\frac{1}{2}}} \left| \frac{d^2E}{dx^2} \right| dx$$

$$\leq \int_{x_{i-\frac{1}{2}}}^{x_{i+\frac{1}{2}}} \left| \frac{d\mathcal{Q}}{dx} \right| dx + \mathcal{O}(\Delta x^3)$$

4.5.1 Determination of the Value of a

The three possible choices for the domain of dependence of $Q_i(x)$ for cell i are shown in Fig. 4.24.

The value of a is determined by effectively minimizing $TV(Q_i(x))$. From (4.45),

$$\frac{dQ_i}{dx} = \begin{cases} \frac{\Delta Q_{i+\frac{1}{2}}}{\Delta x} + \frac{(\Delta Q_{i+\frac{3}{2}} - \Delta Q_{i+\frac{1}{2}})}{\Delta x^2}\left(x - x_{i+\frac{1}{2}}\right) & \text{for } a = 0 \\[2mm] \frac{\Delta Q_{i+\frac{1}{2}}}{\Delta x} + \frac{(\Delta Q_{i+\frac{1}{2}} - \Delta Q_{i-\frac{1}{2}})}{\Delta x^2}\left(x - x_{i+\frac{1}{2}}\right) & \text{for } a = 1 \\[2mm] \frac{\Delta Q_{i-\frac{1}{2}}}{\Delta x} + \frac{(\Delta Q_{i+\frac{1}{2}} - \Delta Q_{i-\frac{1}{2}})}{\Delta x^2}\left(x - x_{i-\frac{1}{2}}\right) & \text{for } a = 1 \\[2mm] \frac{\Delta Q_{i-\frac{1}{2}}}{\Delta x} + \frac{(\Delta Q_{i-\frac{1}{2}} - \Delta Q_{i-\frac{3}{2}})}{\Delta x^2}\left(x - x_{i-\frac{1}{2}}\right) & \text{for } a = 2 \end{cases} \qquad (4.46)$$

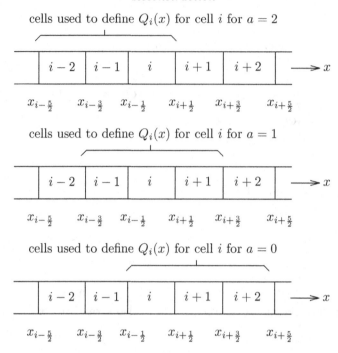

Fig. 4.24. Choices for cells to define $Q_i(x)$

Note that the two expressions for the case $a = 1$ are equivalent. We rewrite the above as

$$\frac{dQ_i}{dx} = \frac{\alpha}{\Delta x} + \frac{\beta}{\Delta x^2}\left(x - x_{i+k-\frac{1}{2}}\right) \qquad (4.47)$$

where α, β, and k are defined in Table 4.3.

Table 4.3. *Derivative of Reconstructed Function in Cell i*

a	α	β	k
0	$\Delta Q_{i+\frac{1}{2}}$	$\Delta Q_{i+\frac{3}{2}} - \Delta Q_{i+\frac{1}{2}}$	1
1	$\Delta Q_{i+\frac{1}{2}}$	$\Delta Q_{i+\frac{1}{2}} - \Delta Q_{i-\frac{1}{2}}$	1
1	$\Delta Q_{i-\frac{1}{2}}$	$\Delta Q_{i+\frac{1}{2}} - \Delta Q_{i-\frac{1}{2}}$	0
2	$\Delta Q_{i-\frac{1}{2}}$	$\Delta Q_{i-\frac{1}{2}} - \Delta Q_{i-\frac{3}{2}}$	0

Therefore,

$$TV(Q_i(x)) = \int_{x_{i-\frac{1}{2}}}^{x_{i+\frac{1}{2}}} \left|\frac{dQ_i}{dx}\right| dx$$

$$\leq \int_{x_{i-\frac{1}{2}}}^{x_{i+\frac{1}{2}}} \frac{|\alpha|}{\Delta x} dx + \int_{x_{i-\frac{1}{2}}}^{x_{i+\frac{1}{2}}} \frac{|\beta|}{\Delta x^2} \left|\left(x - x_{i+k-\frac{1}{2}}\right)\right| dx$$

$$= |\alpha| + \tfrac{1}{2}|\beta| \tag{4.48}$$

Following Harten *et al.* (1987a), we select a to minimize α and β in turn. The algorithm is indicated in Table 4.4.

Table 4.4. *Algorithm for Determining Value of a*

1st criterion	2nd criterion		k	a
$|\Delta Q_{i+\frac{1}{2}}| \leq |\Delta Q_{i-\frac{1}{2}}|$	$|\Delta Q_{i+\frac{3}{2}} - \Delta Q_{i+\frac{1}{2}}| \leq |\Delta Q_{i+\frac{1}{2}} - \Delta Q_{i-\frac{1}{2}}|$		1	0
	$|\Delta Q_{i+\frac{3}{2}} - \Delta Q_{i+\frac{1}{2}}| > |\Delta Q_{i+\frac{1}{2}} - \Delta Q_{i-\frac{1}{2}}|$		1	1
$|\Delta Q_{i+\frac{1}{2}}| > |\Delta Q_{i-\frac{1}{2}}|$	$|\Delta Q_{i+\frac{1}{2}} - \Delta Q_{i-\frac{1}{2}}| \leq |\Delta Q_{i-\frac{1}{2}} - \Delta Q_{i-\frac{3}{2}}|$		0	1
	$|\Delta Q_{i+\frac{1}{2}} - \Delta Q_{i-\frac{1}{2}}| > |\Delta Q_{i-\frac{1}{2}} - \Delta Q_{i-\frac{3}{2}}|$		0	2

An alternate approach for Essentially Non-Oscillatory reconstruction was developed by Liu *et al.* (1994). Instead of selecting the value of a that yields the reconstruction with the minimum oscillation, all possible reconstructions for a given order accuracy are generated and a weighted sum is used as the actual reconstruction. This method is denoted Weighted Essentially Non-Oscillatory reconstruction.

4.5.2 Results

The reconstruction for the sine wave with ten cells is shown in Fig. 4.25. The reconstructed values at the left and right sides of each face are similar to Fig. 4.5. The reconstruction of the discontinuity (4.18) with ten cells is shown in Fig. 4.26. The correct values are reconstructed at all faces.

Exercises

4.1   Prove (4.19) is second-order accurate if $\kappa = -1$.

SOLUTION

Equation (4.19) becomes

$$Q^l_{i+\frac{1}{2}} = Q_i + \tfrac{1}{2}\Delta Q_{i-\frac{1}{2}}$$

$$Q^r_{i-\frac{1}{2}} = Q_i - \tfrac{1}{2}\Delta Q_{i+\frac{1}{2}}$$

Consider the primitive function

$$I(x) = \int_{x_{i-\frac{3}{2}}}^{x} Q \, dx \quad \text{for} \quad x_{i-\frac{3}{2}} \leq x \leq x_{i+\frac{1}{2}}$$

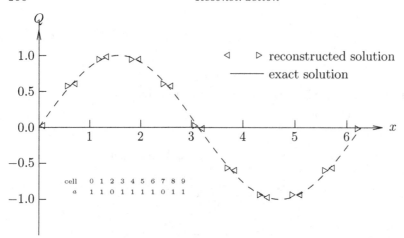

Fig. 4.25. Reconstruction of sine using ten cells

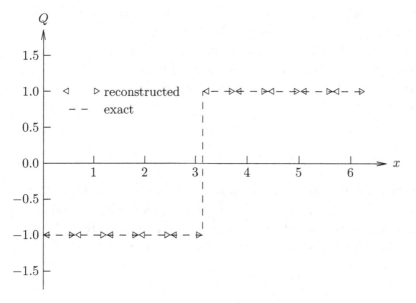

Fig. 4.26. Reconstruction using ten cells

where Δx is constant. Note that the range of the function is less than (4.2). Then

$$
\begin{aligned}
I(x_{i-\frac{3}{2}}) &= 0 \\
I(x_{i-\frac{1}{2}}) &= \Delta x \, Q_{i-1} \\
I(x_{i+\frac{1}{2}}) &= \Delta x \, (Q_{i-1} + Q_i)
\end{aligned}
$$

There is a unique second-order polynomial that interpolates $I(x)$ at the three points $x_{i-\frac{3}{2}}$, $x_{i-\frac{1}{2}}$, and $x_{i+\frac{1}{2}}$. The polynomial may be obtained using Newton's formula

(Mathews and Fink, 1998)

$$P(x) = a_0 + a_1(x - x_{i-\frac{3}{2}}) + a_2(x - x_{i-\frac{3}{2}})(x - x_{i-\frac{1}{2}})$$

The coefficients are defined by

$$
\begin{aligned}
a_0 &= I[x_{i-\frac{3}{2}}] \\
a_1 &= I[x_{i-\frac{3}{2}}, x_{i-\frac{1}{2}}] \\
a_2 &= I[x_{i-\frac{3}{2}}, x_{i-\frac{1}{2}}, x_{i+\frac{1}{2}}]
\end{aligned}
$$

where $I[x_{i-\frac{3}{2}}]$, $I[x_{i-\frac{3}{2}}, x_{i-\frac{1}{2}}]$, and $I[x_{i-\frac{3}{2}}, x_{i-\frac{1}{2}}, x_{i+\frac{1}{2}}]$ are defined by (4.6). Thus,

$$
\begin{aligned}
a_0 &= 0 \\
a_1 &= Q_{i-1} \\
a_2 &= (2\Delta x)^{-1} \Delta Q_{i-\frac{1}{2}}
\end{aligned}
$$

Then

$$I(x) = P(x) + E(x) \quad \text{for} \quad x_{i-\frac{3}{2}} \le x \le x_{i+\frac{1}{2}}$$

where $E(x)$ is the error defined by

$$E(x) = \frac{(x - x_{i-\frac{3}{2}})(x - x_{i-\frac{1}{2}})(x - x_{i+\frac{1}{2}})}{3!} \left. \frac{d^3 I}{dx^3} \right|_{x = \hat{x}}$$

where $\hat{x}$ depends on x and $x_{i-\frac{3}{2}} \le \hat{x} \le x_{i+\frac{1}{2}}$. The reconstruction function for $\mathcal{Q}$ in cell i is denoted as $Q_i(x)$ and is defined by

$$Q_i(x) = \frac{dP}{dx} \quad \text{for} \quad x_{i-\frac{1}{2}} \le x \le x_{i+\frac{1}{2}}$$

and thus

$$Q_i(x) = \mathcal{Q}(x) + \mathcal{O}(\Delta x^2)$$

Now

$$Q_i(x) = Q_i + \frac{\Delta Q_{i-\frac{1}{2}}}{\Delta x} \xi$$

where $\xi = x - x_i$. Thus,

$$Q^l_{i+\frac{1}{2}} = Q_i + \tfrac{1}{2} \Delta Q_{i-\frac{1}{2}}$$

is a second-order accurate reconstruction of $\mathcal{Q}$ to the left face of $x_{i+\frac{1}{2}}$ using cells $i-1$ and i. Similarly, evaluating $Q(x)$ at $x_{i-\frac{3}{2}}$,

$$Q^r_{i-\frac{3}{2}} = Q_{i-1} - \tfrac{1}{2} \Delta Q_{i-\frac{1}{2}}$$

is a second-order accurate reconstruction of $\mathcal{Q}$ to the right face of $x_{i-\frac{3}{2}}$ using cells $i-1$ and i. Letting $i \to i+1$,

$$Q^r_{i-\frac{1}{2}} = Q_i - \tfrac{1}{2} \Delta Q_{i+\frac{1}{2}}$$

is a second-order accurate reconstruction of $\mathcal{Q}$ to the right face of $x_{i-\frac{1}{2}}$ using cells i and $i+1$.

4.2   Verify that the reconstruction $Q_i(x)$ in (4.14) yields

$$Q_i = \frac{1}{V_i} \int_{V_i} Q_i(x) \, dx$$

4.3   Prove (4.19) is second-order accurate if $\kappa = +1$.

Equation (4.19) becomes

$$
\begin{aligned}
Q^l_{i+\frac{1}{2}} &= Q_i + \tfrac{1}{2}\Delta Q_{i+\frac{1}{2}} \\
Q^r_{i-\frac{1}{2}} &= Q_i - \tfrac{1}{2}\Delta Q_{i-\frac{1}{2}}
\end{aligned}
$$

Using the same primitive function as in Exercise 4.1 and letting $i \to i+1$,

$$
Q^l_{i+\frac{1}{2}} = Q_i + \tfrac{1}{2}\Delta Q_{i+\frac{1}{2}}
$$

Using the primitive function again,

$$
Q^r_{i-\frac{1}{2}} = Q_i - \tfrac{1}{2}\Delta Q_{i-\frac{1}{2}}
$$

4.4   Prove (4.19) is second-order accurate if $\kappa = 0$.

4.5   Prove the following reconstruction is first-order accurate:

$$
\begin{aligned}
Q^l_{i+\frac{1}{2}} &= Q_i \\
Q^r_{i-\frac{1}{2}} &= Q_i
\end{aligned}
$$

The solution is analogous to the derivation in Section 4.2. We define the primitive function

$$
I(x) = \int_{x-\frac{1}{2}}^{x} Q\, dx \quad \text{for} \quad x_{i-\frac{1}{2}} \le x \le x_{i+\frac{1}{2}}
$$

Then

$$
\begin{aligned}
I(x_{i-\frac{1}{2}}) &= 0 \\
I(x_{i+\frac{1}{2}}) &= \Delta x\, Q_i
\end{aligned}
$$

The first-order polynomial $P(x)$ that interpolates $I(x)$ at the two points $x_{i-\frac{1}{2}}$ and $x_{i\frac{1}{2}}$ is

$$
P(x) = a_0 + a_1 \left(x - x_{i-\frac{1}{2}} \right)
$$

where

$$
\begin{aligned}
a_0 &= I[x_{i-\frac{1}{2}}] \\
a_1 &= I[x_{i-\frac{1}{2}}, x_{i+\frac{1}{2}}]
\end{aligned}
$$

Now

$$
I(x) = P(x) + E(x) \quad \text{for} \quad x_{i-\frac{1}{2}} \le x \le x_{i+\frac{1}{2}}
$$

where $E(x)$ is defined by

$$
E(x) = \frac{(x - x_{i-\frac{1}{2}})(x - x_{i+\frac{1}{2}})}{2!} \left. \frac{d^2 I}{dx^2} \right|_{x=\hat{x}}
$$

where $x_{i-\frac{1}{2}} \leq \hat{x} \leq x_{i+\frac{1}{2}}$. Thus,

$$
\begin{aligned}
Q(x) &= \frac{dP}{dx} \\
&= a_1 \\
&= Q_i
\end{aligned}
$$

and

$$
\begin{aligned}
Q(x) &= \frac{dI}{dx} - \frac{dE}{dx} \\
&= \mathcal{Q}(x) + \mathcal{O}(\Delta x)
\end{aligned}
$$

4.6 Derive Equation (4.29).

4.7 Prove that the limiter in Section 4.4 for $\kappa = -1$ and 0 is equivalent to

$$
\begin{aligned}
\widehat{\Delta Q}_{i-\frac{1}{2}} &= \text{minmod}\left(\Delta Q_{i-\frac{1}{2}}, b\Delta Q_{i+\frac{1}{2}}\right) \\
\widehat{\Delta Q}_{i+\frac{1}{2}} &= \text{minmod}\left(b\Delta Q_{i-\frac{1}{2}}, \Delta Q_{i+\frac{1}{2}}\right)
\end{aligned}
$$

where

$$
\text{minmod}\,(x,y) = \begin{cases} x & \text{if } |x| \leq |y| \text{ where } x \text{ and } y \text{ have the same sign} \\ y & \text{if } |x| > |y| \text{ where } x \text{ and } y \text{ have the same sign} \\ 0 & \text{where } x \text{ and } y \text{ have opposite signs} \end{cases}
$$

SOLUTION

Consider the expression

$$
\widehat{\Delta Q}_{i-\frac{1}{2}} = \text{minmod}\left(\Delta Q_{i-\frac{1}{2}}, b\Delta Q_{i+\frac{1}{2}}\right)
$$

For $\Delta Q_{i+\frac{1}{2}} \geq 0$ and $\Delta Q_{i-\frac{1}{2}} \geq 0$,

$$
\text{minmod}\left(\Delta Q_{i-\frac{1}{2}}, b\Delta Q_{i+\frac{1}{2}}\right) = \begin{cases} \Delta Q_{i-\frac{1}{2}} & \text{if } \Delta Q_{i-\frac{1}{2}} \leq b\Delta Q_{i+\frac{1}{2}} \\ b\Delta Q_{i+\frac{1}{2}} & \text{if } \Delta Q_{i-\frac{1}{2}} > b\Delta Q_{i+\frac{1}{2}} \end{cases}
$$

which agrees with the result for Case 1. For $\Delta Q_{i+\frac{1}{2}} \geq 0$ and $\Delta Q_{i-\frac{1}{2}} \leq 0$, and for $\Delta Q_{i+\frac{1}{2}} \leq 0$ and $\Delta Q_{i-\frac{1}{2}} \geq 0$,

$$
\text{minmod}\left(\Delta Q_{i-\frac{1}{2}}, b\Delta Q_{i+\frac{1}{2}}\right) = 0
$$

which agrees with the results for Cases 2 and 4, respectively. Finally, for $\Delta Q_{i+\frac{1}{2}} \leq 0$ and $\Delta Q_{i-\frac{1}{2}} \leq 0$,

$$
\text{minmod}\left(\Delta Q_{i-\frac{1}{2}}, b\Delta Q_{i+\frac{1}{2}}\right) = \begin{cases} \Delta Q_{i-\frac{1}{2}} & \text{if } \Delta Q_{i-\frac{1}{2}} \geq b\Delta Q_{i+\frac{1}{2}} \\ b\Delta Q_{i+\frac{1}{2}} & \text{if } \Delta Q_{i-\frac{1}{2}} < b\Delta Q_{i+\frac{1}{2}} \end{cases}
$$

which agrees with the result for Case 3. A similar analysis holds for $\widehat{\Delta Q}_{i+\frac{1}{2}}$.

4.8 Develop the third-order reconstruction analogous to Equation (4.14) assuming an arbitrary mesh spacing Δx_i.

4.9   Show that the two expressions for $a = 1$ in Equation (4.46) are equivalent.

The first expression for $a = 1$ is

$$\frac{dQ_i}{dx} = \frac{\Delta Q_{i+\frac{1}{2}}}{\Delta x} + \frac{(\Delta Q_{i+\frac{1}{2}} - \Delta Q_{i-\frac{1}{2}})}{\Delta x^2}\left(x - x_{i+\frac{1}{2}}\right)$$

Using

$$\begin{aligned}
x - x_{i+\frac{1}{2}} &= x - x_{i-\frac{1}{2}} - (x_{i+\frac{1}{2}} - x_{i-\frac{1}{2}}) \\
&= x - x_{i-\frac{1}{2}} - \Delta x \qquad\qquad\text{(E4.1)}
\end{aligned}$$

the second expression for $a = 1$ is obtained.

4.10   Prove that the algorithm in Table 4.4 for determining the value of a is equivalent to the following iterative scheme (Harten *et al.*, 1987a; see also Chu, 1997) for determining the leftmost cell of the three cells used to reconstruct $Q_i(x)$ in cell i:

(a) Denote the index of the leftmost cell used in the reconstruction of cell i as $\hat{i}_k(i)$, where k is the iteration number beginning with $k = 1$.

(b) Since the reconstruction must include cell i, set $\hat{i}_1(i) = i$.

(c) Select the next iterate for the index of the leftmost cell according to

$$\hat{i}_{k+1}(i) = \begin{cases} \hat{i}_k(i) - 1 & \text{if } |I[x_{\hat{i}_k(i)-\frac{3}{2}}, x_{\hat{i}_k(i)-\frac{1}{2}}, \ldots, x_{\hat{i}_k(i)-\frac{1}{2}+k}]| < \\ & \qquad |I[x_{\hat{i}_k(i)-\frac{1}{2}}, x_{\hat{i}_k(i)+\frac{1}{2}}, \ldots, x_{\hat{i}_k(i)+\frac{1}{2}+k}]| \\ \hat{i}_k(i) & \text{otherwise} \end{cases}$$

for $k = 1, 2$.

5

Godunov Methods

If I would have read Lax's paper [Lax 1959] a year earlier, "Godunov's Scheme" would never have been created.

Sergei Godunov (1999)

5.1 Introduction

The reconstruction methods described in Chapter 4 inherently give rise to a discontinuity in the reconstructed functions at the cell faces. For smooth functions, the discontinuity may be slight as illustrated in Figs. 4.5, 4.20, and 4.25. For discontinuous functions, the difference between the left and right states can be of the same order as the function value itself as shown in Figs. 4.21 and 4.26. This discontinuity in the reconstructed functions is analogous to the Riemann problem described in Section 2.9 and therefore suggests the development of a flux algorithm based on the solution of the Riemann problem. The original concept was developed by Godunov (1959). The class of flux algorithms based on exact or approximate solutions of the Riemann problem (or some generalization thereof) are known as *Godunov, Riemann* or *Flux Difference Splitting Methods* (Hirsch, 1988; Van Leer, 1997). In this chapter, we present three different methods. Additional algorithms are described, for example, in Toro (1997).

5.2 Godunov's Method

Godunov (1959) introduced the concept of utilizing the solution of the local Riemann problem at each cell face as the basis for determining the flux $F_{i \pm \frac{1}{2}}$

in the integral form of the Euler equations (3.10):

$$Q_i^{n+1} = Q_i^n - \frac{1}{\Delta x} \int_{t^n}^{t^{n+1}} \left(F_{i+\frac{1}{2}} - F_{i-\frac{1}{2}} \right) dt$$

5.2.1 Algorithm

Consider the solution Q_i for $i = 1, \ldots, M$ at time t^n. A typical component of Q_i is illustrated in Fig. 5.1. In general, there is a discontinuity in Q_i at each cell face.

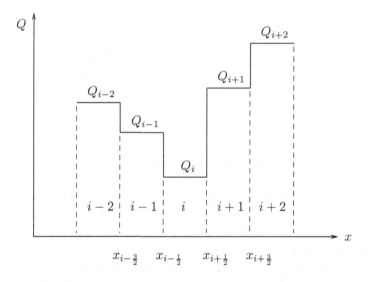

Fig. 5.1. A typical component of Q_i at time t^n

The flux $F_{i+\frac{1}{2}}$ is determined from the solution of the local general Riemann problem (Section 2.9) at $x_{i+\frac{1}{2}}$ (Fig. 5.2). The left and right states $Q_{i+\frac{1}{2}}^l$ and $Q_{i+\frac{1}{2}}^r$ at $x_{i+\frac{1}{2}}$ may be taken to be

$$\begin{aligned} Q_{i+\frac{1}{2}}^l &= Q_i \\ Q_{i+\frac{1}{2}}^r &= Q_{i+1} \end{aligned} \tag{5.1}$$

which corresponds to a first-order accurate reconstruction of Q to the cell face.† The solution to the general Riemann problem at $x_{i+\frac{1}{2}}$ for $t > t^n$ depends only on $(x - x_{i+\frac{1}{2}})/(t - t^n)$ and not on x or t separately. Consequently,

---

† Alternately, a higher order accurate reconstruction can be performed using the methods of Chapter 4.

the solution for $\mathcal{Q}$ at $x_{i+\frac{1}{2}}$ is independent of time, and therefore

$$\int_{t^n}^{t^{n+1}} F_{i+\frac{1}{2}} dt = F_{i+\frac{1}{2}} \Delta t = \mathcal{F}(\mathcal{Q}_{i+\frac{1}{2}}^R) \Delta t \qquad (5.2)$$

where $\mathcal{Q}_{i+\frac{1}{2}}^R$ is the solution of the general Riemann problem at $x_{i+\frac{1}{2}}$ using the reconstructed values $Q_{i+\frac{1}{2}}^l$ and $Q_{i+\frac{1}{2}}^r$. A similar result holds for $F_{i-\frac{1}{2}}$. Thus,

$$Q_i^{n+1} = Q_i^n - \frac{\Delta t}{\Delta x} \left(\mathcal{F}(\mathcal{Q}_{i+\frac{1}{2}}^R) - \mathcal{F}(\mathcal{Q}_{i-\frac{1}{2}}^R) \right) \qquad (5.3)$$

This is Godunov's Second Method.[1]

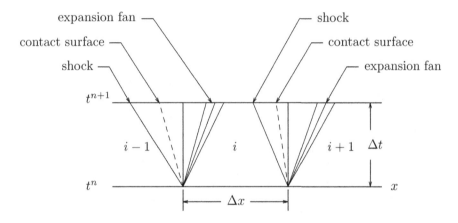

Fig. 5.2. Possible flow structure for Godunov's Second Method

5.2.2 Stability

Intuitively, the time step Δt must be chosen to insure that the rightmost wave emanating from the general Riemann problem at the left face does not intersect the right face, and vice versa. This implies $\Delta t = \Delta t_{CFL}$, where

$$\Delta t_{CFL} = \min_i \frac{\Delta x}{c_{\max}} \qquad (5.4)$$

where $c_{\max}$ is the maximum absolute wave speed of the waves entering cell i and $\min_i$ indicates the minimum value over all cells. The linear approximation to (5.4) is

$$\Delta t_{CFL} \approx \min_i \left(\min_j \frac{\Delta x}{|\lambda_j|} \right) \qquad (5.5)$$

where $\lambda_j, j = 1, 2, 3$ are given by (2.12) and the minimum is taken over all cells. Experience indicates that this criterion may be an overestimate depending on the particular choice of parameters for the algorithm (*e.g.,* reconstruction method) and flow, and consequently a more conservative estimate is employed,

$$\Delta t = \mathcal{C}\Delta t_{CFL} \tag{5.6}$$

where the Courant number $\mathcal{C}$ is less than unity.

5.2.3 Accuracy, Consistency, and Convergence

We now consider the problem described in Section 2.8. The initial condition is defined by (2.92) with $\epsilon = 0.1$, and the domain is $0 < \kappa x < 2\pi$. The first-order reconstruction (5.1) is employed. The norm, defined by (3.102), is evaluated at $\kappa a_o t = 7$, which is prior to the shock formation. The time step Δt is determined by

$$\Delta t = \mathcal{C}\Delta t_{CFL} \tag{5.7}$$

where the Courant number $\mathcal{C} = 0.46$ and Δt_{CFL} is calculated according to

$$\Delta t_{CFL} = \min_i \left(\min_j \frac{\Delta x}{|\lambda_j|} \right) \tag{5.8}$$

using the initial condition. The time step Δt is held in fixed ratio to the grid spacing Δx, and therefore Δt and Δx are decreased by the same ratio. The exact solution $Q_i^{n,e}$ is obtained from (2.85) to (2.88).

The convergence is displayed in Fig. 5.3. The solution converges linearly[†] to the exact solution for sufficiently small Δt. The linear convergence is a direct consequence of the linear reconstruction (5.1). The computed result (using 100 cells) and exact solution are displayed in Fig. 5.4. The significant error in amplitude is attributable to the dissipative nature of the first-order reconstruction (5.1). The speed of the disturbance is accurately predicted, however. The error in amplitude is substantially reduced by using a second-order reconstruction as indicated in Fig. 5.5, where the results are shown using (4.20) with $\kappa = 0$ and the minmod limiter, together with results for the first-order reconstruction, for $\kappa a_o t = 7$ and $\mathcal{C} = 0.18$.

---

† The linear line in Fig. 5.3 is

$$\|Q_i^{n,e} - Q_i^n\|_{\kappa a_o \Delta t} = \left(\frac{\kappa a_o \Delta t}{5 \times 10^{-4}} \right) \|Q_i^{n,e} - Q_i^n\|_{\kappa a_o \Delta t = 5 \times 10^{-4}}$$

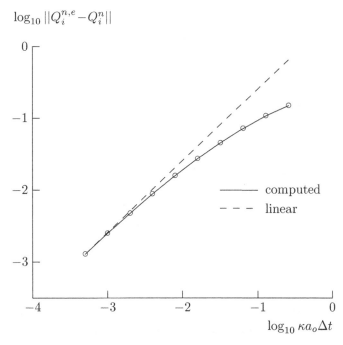

Fig. 5.3. Convergence for Godunov's Second Method using first-order reconstruction

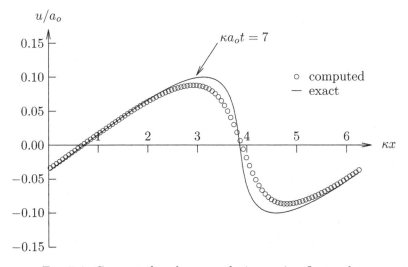

Fig. 5.4. Computed and exact solutions using first-order reconstruction

5.3 Roe's Method

Roe (1981,1986) developed an algorithm based on an exact solution to an approximation of the generalized Riemann problem. Unlike Godunov's Sec-

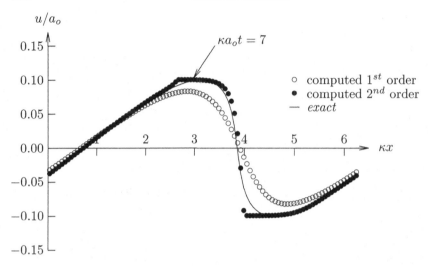

Fig. 5.5. Computed and exact solutions using first- and second-order reconstructions

ond Method, Roe's scheme does not require an iterative procedure to find the flux‡.

5.3.1 Algorithm

Consider the Euler equations (2.6) in nonconservative differential form,

$$\frac{\partial \mathcal{Q}}{\partial t} + \mathcal{A}\frac{\partial \mathcal{Q}}{\partial x} = 0 \qquad (5.9)$$

where $\mathcal{A}$ is the Jacobian (2.11)

$$\mathcal{A}(\mathcal{Q}) = \left\{ \begin{array}{ccc} 0 & 1 & 0 \\ (\gamma-3)\,u^2/2 & (3-\gamma)\,u & \gamma-1 \\ -Hu+(\gamma-1)u^3/2 & H-(\gamma-1)u^2 & \gamma u \end{array} \right\} \qquad (5.10)$$

and $H = e + p/\rho$ is the total enthalpy. Roe sought a solution of the general Riemann problem using an *approximate* form of the Euler equations

$$\frac{\partial \mathcal{Q}}{\partial t} + \tilde{\mathcal{A}}(\mathcal{Q}_l, \mathcal{Q}_r)\frac{\partial \mathcal{Q}}{\partial x} = 0 \qquad (5.11)$$

where $\tilde{\mathcal{A}}(\mathcal{Q}_l, \mathcal{Q}_r)$ depends on the left and right states $\mathcal{Q}_l$ and $\mathcal{Q}_r$ of the general Riemann problem (Fig. 5.6) and is assumed constant.

---

‡ Recall that the solution of the General Riemann Problem requires an iteration to find the contact pressure except in the case of expansion-expansion.

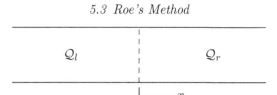

Fig. 5.6. Initial condition for General Riemann Problem

The matrix $\tilde{A}$ is an approximation to the matrix exact A and satisfies the following four properties (Roe, 1981)

(i) $\tilde{A}$ provides a linear mapping from the vector space of Q to the vector space of F.
(ii) $\tilde{A}(Q_l, Q_r) \to A(Q)$ as $Q_l \to Q_r \to Q$.
(iii) For any Q_l and Q_r, $\tilde{A}(Q_l, Q_r) \times (Q_l - Q_r) \equiv F_l - F_r$.
(iv) The eigenvectors of $\tilde{A}(Q_l, Q_r)$ are linearly independent.

Property (i) implies that the individual components of the vector $\tilde{A}Q$ have the same units as the corresponding components of the vector F. Property (ii) implies that the solution to (5.11) is a close approximation to the solution of (5.9) for smooth Q. Property (iii) requires satisfaction of the exact jump conditions across a shock wave. Property (iv) is consistent with the assumed hyperbolic character of (5.11), *i.e.*, $\tilde{A}(Q_l, Q_r)$ is diagonalizable with real, distinct eigenvalues and linearly independent eigenvectors.

In order to determine the matrix $\tilde{A}$, several identities are needed. Consider the arbitrary piecewise constant functions f and g that have left and right states indicated by f_l, g_l and f_r, g_r, respectively. The following identities can be proven (Exercise 5.1):

$$
\begin{aligned}
\Delta(f + g) &= \Delta f + \Delta g \\
\Delta(fg) &= \bar{f}\Delta g + \Delta f \bar{g} \\
\Delta(1/f) &= -\Delta f / \hat{f}^2
\end{aligned}
\tag{5.12}
$$

where $\Delta f = f_l - f_r$ and similarly for Δg, and

$$
\begin{aligned}
\bar{f} &\equiv \tfrac{1}{2}(f_l + f_r) \\
\hat{f} &\equiv \sqrt{f_l f_r}
\end{aligned}
\tag{5.13}
$$

The matrix $\tilde{A}$ is found in the following manner. A *parameterization vector*

$\nu = (\nu_1, \nu_2, \nu_3)^T$ is introduced as

$$\nu = \left\{ \begin{array}{c} \sqrt{\rho} \\ \sqrt{\rho}\,u \\ \sqrt{\rho}\,H \end{array} \right\} \tag{5.14}$$

and thus $\mathcal{Q}$ and $\mathcal{F}$ are

$$\mathcal{Q} = \left\{ \begin{array}{c} \nu_1^2 \\ \nu_1\nu_2 \\ \nu_1\nu_3/\gamma + (\gamma-1)\nu_2^2/2\gamma \end{array} \right\} \tag{5.15}$$

$$\mathcal{F} = \left\{ \begin{array}{c} \nu_1\nu_2 \\ (\gamma-1)\nu_1\nu_3/\gamma + (\gamma+1)\,\nu_2^2/2\gamma \\ \nu_2\nu_3 \end{array} \right\} \tag{5.16}$$

Since $\mathcal{Q}$ and $\mathcal{F}$ are quadratic in the elements of ν, it is possible to find matrices B and C such that

$$\Delta\mathcal{Q} = B\Delta\nu \tag{5.17}$$
$$\Delta\mathcal{F} = C\Delta\nu \tag{5.18}$$

where $\Delta\mathcal{Q} = \mathcal{Q}_l - \mathcal{Q}_r$, and so on (Exercise 5.2).

Therefore

$$\Delta\mathcal{F} = \tilde{A}\Delta\mathcal{Q}$$
$$C\Delta\nu = \tilde{A}B\Delta\nu$$
$$\Delta\nu = C^{-1}\tilde{A}B\Delta\nu \tag{5.19}$$

and thus $\tilde{A} = CB^{-1}$.

Consequently, the determination of the matrix $\tilde{A}$ reduces to the problem of finding B and C. It is straightforward to show (Exercise 5.3)

$$B = \left\{ \begin{array}{ccc} 2\bar{\nu}_1 & 0 & 0 \\ \bar{\nu}_2 & \bar{\nu}_1 & 0 \\ \bar{\nu}_3/\gamma & (\gamma-1)\bar{\nu}_2/\gamma & \bar{\nu}_1/\gamma \end{array} \right\}$$

$$C = \left\{ \begin{array}{ccc} \bar{\nu}_2 & \bar{\nu}_1 & 0 \\ (\gamma-1)\bar{\nu}_3/\gamma & (\gamma+1)\bar{\nu}_2/\gamma & (\gamma-1)\bar{\nu}_1/\gamma \\ 0 & \bar{\nu}_3 & \bar{\nu}_2 \end{array} \right\} \tag{5.20}$$

and thus (Exercise 5.4)

$$\tilde{A} = \left\{ \begin{array}{ccc} 0 & 1 & 0 \\ (\gamma-3)\tilde{u}^2/2 & (3-\gamma)\tilde{u} & (\gamma-1) \\ -\tilde{H}\tilde{u}+(\gamma-1)\tilde{u}^3/2 & \tilde{H}-(\gamma-1)\tilde{u}^2 & \gamma\tilde{u} \end{array} \right\} \tag{5.21}$$

where

$$\tilde{u} \equiv \frac{\bar{\nu}_2}{\bar{\nu}_1} = \frac{\sqrt{\rho_l}u_l + \sqrt{\rho_r}u_r}{\sqrt{\rho_l} + \sqrt{\rho_r}}$$

$$\tilde{H} \equiv \frac{\bar{\nu}_3}{\bar{\nu}_1} = \frac{\sqrt{\rho_l}H_l + \sqrt{\rho_r}H_r}{\sqrt{\rho_l} + \sqrt{\rho_r}} \tag{5.22}$$

The quantities $\tilde{u}$ and $\tilde{H}$ are the *Roe-averaged velocity* and *Roe-averaged total enthalpy*, respectively. The matrix $\tilde{A}(Q_l, Q_r)$ is the *Roe matrix*.

Clearly, $\tilde{A}$ satisfies Property (i). By comparison with (5.10), $\tilde{A}$ satisfies Property (ii), since $\tilde{u} \to u$ and $\tilde{H} \to H$ as $Q_l \to Q_r \to Q$. Property (iii) is satisfied by construction. Finally, the eigenvalues $\tilde{\lambda}_i$ and the right eigenvectors $\tilde{e}_i$ of $\tilde{A}$ may be found directly (Exercise 5.5),

$$\tilde{\lambda}_1 = \tilde{u}, \qquad \tilde{\lambda}_2 = \tilde{u} + \tilde{a}, \qquad \tilde{\lambda}_3 = \tilde{u} - \tilde{a} \tag{5.23}$$

$$\tilde{e}_1 = \left\{ \begin{array}{c} 1 \\ \tilde{u} \\ \frac{1}{2}\tilde{u}^2 \end{array} \right\}, \ \tilde{e}_2 = \left\{ \begin{array}{c} 1 \\ \tilde{u} + \tilde{a} \\ \tilde{H} + \tilde{u}\tilde{a} \end{array} \right\}, \ \tilde{e}_3 = \left\{ \begin{array}{c} 1 \\ \tilde{u} - \tilde{a} \\ \tilde{H} - \tilde{u}\tilde{a} \end{array} \right\} \tag{5.24}$$

where $\tilde{a}$ is the speed of sound based on the Roe-averaged total enthalpy and velocity and is given by

$$\tilde{a} = \sqrt{(\gamma-1)(\tilde{H} - \tfrac{1}{2}\tilde{u}^2)} \tag{5.25}$$

It may be directly verified that the eigenvectors are linearly independent, thus satisfying Property (iv).

The *exact* solution of the *approximate* General Riemann Problem

$$\frac{\partial Q}{\partial t} + \tilde{A}(Q_l, Q_r)\frac{\partial Q}{\partial x} = 0 \tag{5.26}$$

is now sought, where $\tilde{A}(Q_l, Q_r)$ is treated as a constant. The Roe matrix may be diagonalized as

$$\tilde{A}(Q_l, Q_r) = \tilde{S}\tilde{\Lambda}\tilde{S}^{-1} \tag{5.27}$$

where $\tilde{S}$ is matrix of right eigenvectors of $\tilde{A}(Q_l, Q_r)$,

$$\tilde{S} = \left\{ \begin{array}{ccc} 1 & 1 & 1 \\ \tilde{u} & \tilde{u} + \tilde{a} & \tilde{u} - \tilde{a} \\ \frac{1}{2}\tilde{u}^2 & \tilde{H} + \tilde{u}\tilde{a} & \tilde{H} - \tilde{u}\tilde{a} \end{array} \right\} \tag{5.28}$$

and $\tilde{S}^{-1}$ is

$$\tilde{S}^{-1} = \left\{ \begin{array}{ccc} 1-(\gamma-1)\tilde{u}^2/2\tilde{a}^2 & (\gamma-1)\tilde{u}/\tilde{a}^2 & -(\gamma-1)/\tilde{a}^2 \\ (\gamma-1)\tilde{u}^2/4\tilde{a}^2 - \tilde{u}/2\tilde{a} & -(\gamma-1)\tilde{u}/2\tilde{a}^2 + 1/2\tilde{a} & (\gamma-1)/2\tilde{a}^2 \\ (\gamma-1)\tilde{u}^2/4\tilde{a}^2 + \tilde{u}/2\tilde{a} & -(\gamma-1)\tilde{u}/2\tilde{a}^2 - 1/2\tilde{a} & (\gamma-1)/2\tilde{a}^2 \end{array} \right\}$$

(5.29)

Since $\tilde{\mathcal{A}}(\mathcal{Q}_l, \mathcal{Q}_r)$ is treated as a constant, it is possible to multiply (5.26) by $\tilde{S}^{-1}$ to obtain

$$\frac{\partial R}{\partial t} + \tilde{\Lambda}\frac{\partial R}{\partial x} = 0$$

(5.30)

where[2]

$$R \equiv \tilde{S}^{-1}\mathcal{Q} = \left\{ \begin{array}{c} R_1 \\ R_2 \\ R_3 \end{array} \right\}$$

(5.31)

and $\tilde{\Lambda}$ is

$$\tilde{\Lambda} \equiv \left\{ \begin{array}{ccc} \tilde{\lambda}_1 & 0 & 0 \\ 0 & \tilde{\lambda}_2 & 0 \\ 0 & 0 & \tilde{\lambda}_3 \end{array} \right\}$$

(5.32)

The solution of (5.30) is

$$R_1 = \text{constant} \quad \text{on curve } C_1 \text{ defined by } \frac{dx}{dt} = \tilde{\lambda}_1 = \tilde{u}$$

$$R_2 = \text{constant} \quad \text{on curve } C_2 \text{ defined by } \frac{dx}{dt} = \tilde{\lambda}_2 = \tilde{u} + \tilde{a}$$

$$R_3 = \text{constant} \quad \text{on curve } C_2 \text{ defined by } \frac{dx}{dt} = \tilde{\lambda}_3 = \tilde{u} - \tilde{a}$$

C_1, C_2, and C_3 are the *characteristic curves* of (5.30). The solution is illustrated in Fig. 5.7 assuming $0 < \tilde{u} < \tilde{a}$. The solutions for R_1, R_2, and R_3 are shown together in Fig. 5.8.

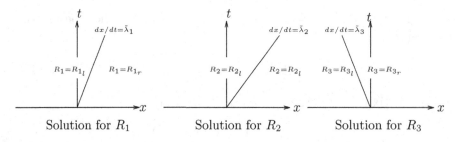

Fig. 5.7. Roe solution to the General Riemann Problem

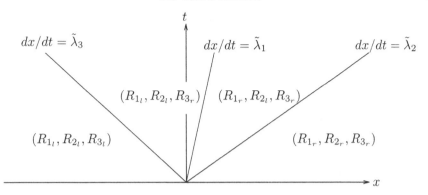

Fig. 5.8. Roe solution to the General Riemann Problem

Consider now the semi-discrete form (3.9) of the Euler equations

$$\frac{dQ_i}{dt} + \frac{\left(F_{i+\frac{1}{2}} - F_{i-\frac{1}{2}}\right)}{\Delta x} = 0$$

The flux $F_{i+\frac{1}{2}}$ given by (4.1) is

$$F_{i+\frac{1}{2}} = \mathcal{F}_{i+\frac{1}{2}} = \tilde{A}Q = \left(\tilde{S}\tilde{\Lambda}\tilde{S}^{-1}\right)\left(\tilde{S}R\right)_{i+\frac{1}{2}} = \tilde{S}\tilde{\Lambda}R_{i+\frac{1}{2}} \qquad (5.33)$$

The individual components of R at $x_{i+\frac{1}{2}}$ (i.e., $x = 0$ in Fig. 5.8) are given by

$$R_k|_{i+1/2}^{n+1} = \tfrac{1}{2}\left(R_{k_l} + R_{k_r}\right) + \tfrac{1}{2}\,\text{sign}(\tilde{\lambda}_k)\left(R_{k_l} - R_{k_r}\right) \qquad (5.34)$$

for $k = 1,2,3$ where

$$\text{sign}(\tilde{\lambda}_l) = \begin{cases} +1 & \text{if } \tilde{\lambda}_k > 0 \\ -1 & \text{if } \tilde{\lambda}_k < 0 \\ \;\;0 & \text{if } \tilde{\lambda}_k = 0 \end{cases} \qquad (5.35)$$

It is straightforward to show that

$$F_{i+\frac{1}{2}} = \tfrac{1}{2}\tilde{S}\tilde{\Lambda}\left(R_l + R_r\right) + \tfrac{1}{2}\tilde{S}|\tilde{\Lambda}|(R_l - R_r) \qquad (5.36)$$

where

$$|\tilde{\Lambda}| = \begin{Bmatrix} |\tilde{\lambda}_1| & 0 & 0 \\ 0 & |\tilde{\lambda}_2| & 0 \\ 0 & 0 & |\tilde{\lambda}_3| \end{Bmatrix} \qquad (5.37)$$

Using (5.31),

$$F_{i+\frac{1}{2}} = \tfrac{1}{2}\left[\tilde{S}\tilde{\Lambda}\tilde{S}^{-1}(Q_l + Q_r) + \tilde{S}|\tilde{\Lambda}|\tilde{S}^{-1}(Q_l - Q_r)\right] \qquad (5.38)$$

Since $\tilde{S}\tilde{\Lambda}\tilde{S}^{-1}\mathcal{Q}_l = \tilde{A}\mathcal{Q}_l$ is the flux based on the flow variables on the left side and similarly for the right side, then

$$F_{i+\frac{1}{2}} = \tfrac{1}{2}\left[F_l + F_r + \tilde{S}|\tilde{\Lambda}|\tilde{S}^{-1}(\mathcal{Q}_l - \mathcal{Q}_r)\right] \tag{5.39}$$

Approximating the left and right states by the reconstructed values

$$\mathcal{Q}_l = Q^l_{i+\frac{1}{2}}$$
$$\mathcal{Q}_r = Q^r_{i+\frac{1}{2}}$$

the flux becomes

$$F_{i+\frac{1}{2}} = \tfrac{1}{2}\left[F_l + F_r + \tilde{S}|\tilde{\Lambda}|\tilde{S}^{-1}(Q^l_{i+\frac{1}{2}} - Q^r_{i+\frac{1}{2}})\right] \tag{5.40}$$

where $F_l = F(Q^l_{i+\frac{1}{2}})$ and $F_r = F(Q^r_{i+\frac{1}{2}})$. This is *Roe's method* for the flux F. It uses an *exact* solution to the *approximate* Riemann Problem (5.26). The flux $F_{i+\frac{1}{2}}$ in (5.40) is taken to be an *approximate* expression for the flux in the *exact* Riemann Problem.

A more convenient form is

$$F_{i+\frac{1}{2}} = \tfrac{1}{2}\left[F_l + F_r + + \sum_{j=1}^{j=3}\alpha_j|\tilde{\lambda}_j|\tilde{e}_j\right] \tag{5.41}$$

where

$$\alpha_1 = \left[1 - \frac{(\gamma-1)}{2}\frac{\tilde{u}^2}{\tilde{a}^2}\right]\Delta\rho + \left[(\gamma-1)\frac{\tilde{u}}{\tilde{a}^2}\right]\Delta\rho u -$$
$$\left[\frac{(\gamma-1)}{\tilde{a}^2}\right]\Delta\rho e$$

$$\alpha_2 = \left[\frac{(\gamma-1)}{4}\frac{\tilde{u}^2}{\tilde{a}^2} - \frac{\tilde{u}}{2\tilde{a}}\right]\Delta\rho + \left[\frac{1}{2\tilde{a}} - \frac{(\gamma-1)}{2}\frac{\tilde{u}}{\tilde{a}^2}\right]\Delta\rho u +$$
$$\left[\frac{(\gamma-1)}{2\tilde{a}^2}\right]\Delta\rho e$$

$$\alpha_3 = \left[\frac{(\gamma-1)}{4}\frac{\tilde{u}^2}{\tilde{a}^2} + \frac{\tilde{u}}{2\tilde{a}}\right]\Delta\rho - \left[\frac{1}{2\tilde{a}} + \frac{(\gamma-1)}{2}\frac{\tilde{u}}{\tilde{a}^2}\right]\Delta\rho u +$$
$$\left[\frac{(\gamma-1)}{2\tilde{a}^2}\right]\Delta\rho e \tag{5.42}$$

and $\Delta\rho = \rho_l - \rho_r$, and so on. The coefficients α_k may be interpreted as the strength of the kth wave that has speed $\tilde{\lambda}_k$.

5.3.2 Stability

We consider the semi-discrete stability of Roe's Method.† The Euler equations (3.9) are

$$\frac{dQ_i}{dt} + \frac{\left(F_{i+\frac{1}{2}} - F_{i-\frac{1}{2}}\right)}{\Delta x} = 0$$

The flux $F_{i+\frac{1}{2}}$ depends on the left and right states,

$$F_{i+\frac{1}{2}} = F(Q^l_{i+\frac{1}{2}}, Q^r_{i+\frac{1}{2}})$$

We linearize $F_{i+\frac{1}{2}}$ as

$$F_{i+\frac{1}{2}} = \tfrac{1}{2}\left[\tilde{A}(Q^l_{i+\frac{1}{2}} + Q^r_{i+\frac{1}{2}}) + |\tilde{A}|(Q^l_{i+\frac{1}{2}} - Q^r_{i+\frac{1}{2}})\right] \qquad (5.43)$$

where

$$|\tilde{A}| = \tilde{S}|\tilde{\Lambda}|\tilde{S}^{-1}$$

and $\tilde{A}$ and $|\tilde{A}|$ are treated as constants.

We rewrite (3.9) as

$$\frac{dQ_i}{dt} = R_i(Q^l_{i+\frac{1}{2}}, Q^r_{i+\frac{1}{2}}, Q^l_{i-\frac{1}{2}}, Q^r_{i-\frac{1}{2}})$$

where

$$R_i = -\frac{\left(F_{i+\frac{1}{2}} - F_{i-\frac{1}{2}}\right)}{\Delta x} \qquad (5.44)$$

For clarity, we define

$$\mathcal{X}_1 = Q^l_{i+\frac{1}{2}}, \quad \mathcal{X}_2 = Q^r_{i+\frac{1}{2}}, \quad \mathcal{X}_3 = Q^l_{i-\frac{1}{2}}, \quad \mathcal{X}_4 = Q^r_{i-\frac{1}{2}}$$

Thus,

$$R_i = R_i(\mathcal{X}_1, \mathcal{X}_2, \mathcal{X}_3, \mathcal{X}_4)$$

We expand R_i in a Taylor series about Q_i, yielding

$$R_i(Q^l_{i+\frac{1}{2}}, Q^r_{i+\frac{1}{2}}, Q^l_{i-\frac{1}{2}}, Q^r_{i-\frac{1}{2}}) = R_i(Q_i, Q_i, Q_i, Q_i)+$$

$$\frac{\partial R_i}{\partial \mathcal{X}_1}(Q^l_{i+\frac{1}{2}} - Q_i) + \frac{\partial R_i}{\partial \mathcal{X}_2}(Q^r_{i+\frac{1}{2}} - Q_i) + \frac{\partial R_i}{\partial \mathcal{X}_3}(Q^l_{i-\frac{1}{2}} - Q_i) + \frac{\partial R_i}{\partial \mathcal{X}_4}(Q^r_{i-\frac{1}{2}} - Q_i)+$$

$$\mathcal{O}(\Delta Q_i^2)$$

where $\partial R_i/\partial \mathcal{X}_j$, $j = 1, \ldots, 4$, are evaluated at $(\mathcal{X}_1, \mathcal{X}_2, \mathcal{X}_3, \mathcal{X}_4) = (Q_i, Q_i, Q_i, Q_i)$.

† A discrete stability analysis of Roe's Method is presented in Chapter 7.

From (5.40),

$$\frac{\partial F_{i+\frac{1}{2}}}{\partial \mathcal{X}_1} = \tfrac{1}{2}\left(\tilde{A} + |\tilde{A}|\right), \quad \frac{\partial F_{i+\frac{1}{2}}}{\partial \mathcal{X}_2} = \tfrac{1}{2}\left(\tilde{A} - |\tilde{A}|\right)$$

$$\frac{\partial F_{i+\frac{1}{2}}}{\partial \mathcal{X}_3} = 0, \quad \frac{\partial F_{i+\frac{1}{2}}}{\partial \mathcal{X}_4} = 0, \quad \frac{\partial F_{i-\frac{1}{2}}}{\partial \mathcal{X}_1} = 0, \quad \frac{\partial F_{i-\frac{1}{2}}}{\partial \mathcal{X}_2} = 0$$

$$\frac{\partial F_{i-\frac{1}{2}}}{\partial \mathcal{X}_3} = \tfrac{1}{2}\left(\tilde{A} + |\tilde{A}|\right), \quad \frac{\partial F_{i-\frac{1}{2}}}{\partial \mathcal{X}_4} = \tfrac{1}{2}\left(\tilde{A} - |\tilde{A}|\right)$$

To proceed further, the spatial reconstruction must be specified. We select the simple first-order reconstruction

$$\begin{aligned} Q^l_{i+\frac{1}{2}} &= Q_i \\ Q^r_{i+\frac{1}{2}} &= Q_{i+1} \end{aligned} \tag{5.45}$$

Moreover,

$$\begin{aligned} \tilde{A} &= A + \mathcal{O}(\Delta Q) \\ |\tilde{A}| &= |A| + \mathcal{O}(\Delta Q) \end{aligned}$$

and thus, to the lowest order,

$$R_i = -\frac{1}{\Delta x}\left[A^- Q_{i+1} + (A^+ - A^-)Q_i - A^+ Q_{i-1}\right] \tag{5.46}$$

where

$$\begin{aligned} A^+ &= \tfrac{1}{2}(A + |A|) &= T\Lambda^+ T^{-1} \\ A^- &= \tfrac{1}{2}(A - |A|) &= T\Lambda^- T^{-1} \end{aligned} \tag{5.47}$$

where

$$\begin{aligned} \Lambda^+ &= \operatorname{diag}\{\max(\lambda_m, 0)\} \\ \Lambda^- &= \operatorname{diag}\{\min(\lambda_m, 0)\} \end{aligned} \tag{5.48}$$

Thus, Λ^+ is a diagonal matrix whose elements $\max(\lambda_m, 0)$ are thus non-negative. Similarly, Λ^- is a diagonal matrix whose elements $\min(\lambda_m, 0)$ are thus nonpositive. Thus, the linearized semi-discrete form of the Euler equations is

$$\frac{dQ_i}{dt} = -\frac{1}{\Delta x}\left[A^- Q_{i+1} + (A^+ - A^-)Q_i - A^+ Q_{i-1}\right] \tag{5.49}$$

For purposes of simplicity, we assume that the flow is periodic in x over a length $L = (M-1)\Delta x$,

$$Q_1 = Q_M$$

and assume M is odd with $M = 2N+1$. Consider $Q(x,t)$ to be a continuous vector function that interpolates Q_i,

$$Q(x_i, t) = Q_i(t)$$

Then the Fourier series (3.30) for $Q(x,t)$ is

$$Q(x,t) = \sum_{l=-N+1}^{l=N} \hat{Q}_k(t) e^{\iota k x}$$

where $\iota = \sqrt{-1}$ and the wavenumber k depends on l according to (3.31):

$$k = \frac{2\pi l}{L}$$

The Fourier coefficients $\hat{Q}_k(t)$ are complex vectors whose subscript k indicates an ordering with respect to the summation index l, i.e., $\hat{Q}_k(t)$ indicates dependence on l (through (3.31)) and on t. Given the values of Q_i at some time t^n, the Fourier coefficients $\hat{Q}_k$ at t^n are obtained from (3.32).

Substituting into (5.49) yields

$$\frac{d\hat{Q}_k}{dt} = G\hat{Q}_k \qquad (5.50)$$

where the amplification matrix G is

$$G = -\frac{1}{\Delta x} \left[e^{\iota k \Delta x} A^- + A^+ - A^- - e^{-\iota k \Delta x} A^+ \right]$$

From (5.47) and (5.48),

$$G = TDT^{-1}$$

where D is the diagonal matrix

$$D = -\frac{1}{\Delta x} \left[e^{\iota k \Delta x} \Lambda^- + \Lambda^+ - \Lambda^- - e^{-\iota k \Delta x} \Lambda^+ \right] \qquad (5.51)$$

Defining

$$\begin{aligned} \lambda_m^+ &= \max(\lambda_m, 0) \\ \lambda_m^- &= \min(\lambda_m, 0) \end{aligned} \qquad (5.52)$$

it is evident that the eigenvalues of G are

$$\lambda_{G_m} = -\frac{1}{\Delta x} \left[e^{\iota k \Delta x} \lambda_m^- + \lambda_m^+ - \lambda_m^- - e^{-\iota k \Delta x} \lambda_m^+ \right] \qquad (5.53)$$

Defining

$$\tilde{Q}_k = T^{-1}\hat{Q}_k$$

Equation (5.50) becomes

$$\frac{d\tilde{Q}_k}{dt} = D\tilde{Q}_k$$

Defining

$$\tilde{Q}_k = \left\{ \begin{array}{c} \tilde{Q}_{k_1} \\ \tilde{Q}_{k_2} \\ \tilde{Q}_{k_3} \end{array} \right\}$$

the solution is

$$\tilde{Q}_{k_m} = \tilde{Q}_{k_m}(0)e^{\lambda_{G_m}t} \quad \text{for} \quad m = 1, 2, 3$$

The condition for stability is therefore

$$\text{Real}(\lambda_{G_m}) \le 0 \tag{5.54}$$

i.e., all of the eigenvalues $\lambda_{G_m}, m = 1, 2, 3$ of the amplification matrix G lie in the left half or imaginary axis of the complex plane. Now

$$\text{Real}(\lambda_{G_m}) = -\frac{1}{\Delta x} \underbrace{(\lambda_m^+ - \lambda_m^-)}_{\ge 0} \underbrace{(1 - \cos k\Delta x)}_{\ge 0}$$

and hence the stability condition (5.54) is satisfied.

5.3.3 Accuracy, Consistency, and Convergence

We now consider the problem described in Section 2.8. The initial condition is defined by (2.92) with $\epsilon = 0.1$ and the domain is $0 < \kappa x < 2\pi$. The first-order reconstruction (5.1) is employed. The norm, defined by (3.102), is evaluated at $\kappa a_o t = 7$, which is prior to the shock formation. The semi-discrete form (3.9) of the Euler equations is employed and the time integration is performed using a second-order Runge-Kutta method (Chapter 7).

The convergence is displayed in Fig. 5.9. The solution converges linearly[†] to the exact solution for sufficiently small Δt. The linear convergence is a direct consequence of the linear reconstruction (5.1). The computed result (using 100 cells) and exact solution are displayed[‡] in Fig. 5.10 for $\kappa a_o t = 7$

---

† The line in Fig. 5.9 is

$$\|Q_i^{n,e} - Q_i^n\|_{\kappa a_o \Delta t} = \left(\frac{\kappa a_o \Delta t}{5 \times 10^{-4}}\right) \|Q_i^{n,e} - Q_i^n\|_{\kappa a_o \Delta t = 5 \times 10^{-4}}$$

‡ More precisely, the timestep $\kappa a_o \Delta t = 0.025$, which corresponds to $\mathcal{C} = 0.45$ at $t = 0$.

and $\mathcal{C} = 0.45$. The significant error in amplitude is attributable to the dissipative nature of the first-order reconstruction (5.1). The speed of the disturbance is accurately predicted, however.

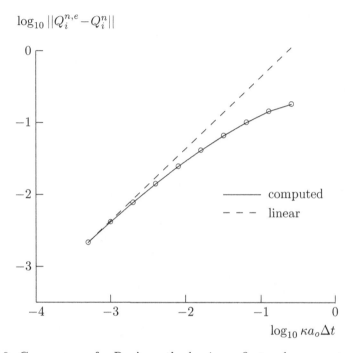

Fig. 5.9. Convergence for Roe's method using a first-order reconstructions

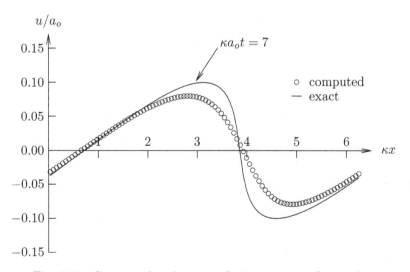

Fig. 5.10. Computed and exact solutions using a first-order reconstruction

The error in amplitude is significantly reduced by using a second-order reconstruction as indicated in Fig. 5.11, where results are shown using (4.20) with $\kappa = 0$ and no limiter together with results for the first-order reconstruction. The convergence for the second-order reconstruction (using second order accurate integration in time) is shown in Fig. 5.12 and displays quadratic convergence.†

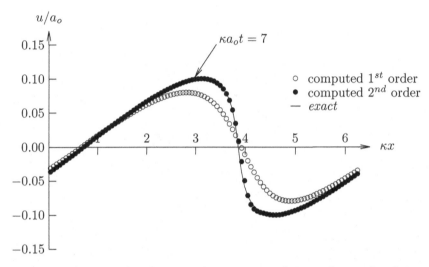

Fig. 5.11. Computed and exact solutions using first- and second-order reconstructions

5.3.4 Entropy Fix

It is interesting to compare the exact solution of the Roe equations (5.11)

$$\frac{\partial Q}{\partial t} + \tilde{A}(Q_l, Q_r)\frac{\partial Q}{\partial x} = 0$$

with the exact solution of the Euler equations (2.1)

$$\frac{\partial Q}{\partial t} + \frac{\partial F}{\partial x} = 0$$

for the Riemann Shock Tube (Section 2.10). The solution to the Roe equations exhibits three waves corresponding to the eigenvalues (5.23). For the

† The line in Fig. 5.12 is

$$\|Q_i^{n,e} - Q_i^n\|_{\kappa a_o \Delta t} = \left(\frac{\kappa a_o \Delta t}{5 \times 10^{-4}}\right)^2 \|Q_i^{n,e} - Q_i^n\|_{\kappa a_o \Delta t = 5 \times 10^{-4}}$$

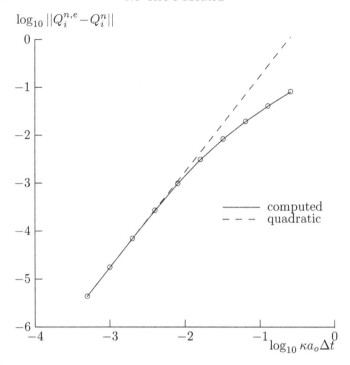

Fig. 5.12. Convergence for Roe's Method using a second-order reconstruction

Riemann Shock Tube,

$$
\begin{aligned}
\tilde{u} &= 0 \\
\tilde{a} &= \left[(\gamma-1)c_p \left(\sqrt{\rho_1}T_1 + \sqrt{\rho_4 T_4} \right) / \left(\sqrt{\rho_1} + \sqrt{\rho_4} \right) \right]^{1/2}
\end{aligned}
$$

and hence $\tilde{a}^2 = (\gamma-1)\tilde{H}$. The Roe eigenvalues are therefore

$$
\begin{aligned}
\tilde{\lambda}_1 &= 0 \\
\tilde{\lambda}_2 &= \tilde{a} \\
\tilde{\lambda}_3 &= -\tilde{a}
\end{aligned}
$$

The three waves divide the $x-t$ domain into four regions (Fig. 5.13). The matrix $\tilde{S}$ is

$$
\tilde{S} = \left\{ \begin{array}{ccc} 1 & 1 & 1 \\ 0 & \tilde{a} & -\tilde{a} \\ 0 & \tilde{a}^2/(\gamma-1) & \tilde{a}^2/(\gamma-1) \end{array} \right\}
$$

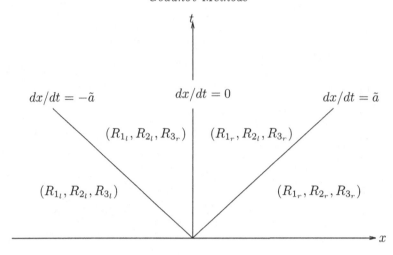

Fig. 5.13. Waves for Riemann Shock Tube

and the vector R is

$$R = \tilde{S}^{-1}\mathcal{Q} = \left\{ \begin{array}{c} \rho - (\gamma-1)\rho e/\tilde{a}^2 \\ \rho u/2\tilde{a} + (\gamma-1)\rho e/2\tilde{a}^2 \\ -\rho u/2\tilde{a} + (\gamma-1)\rho e/2\tilde{a}^2 \end{array} \right\}$$

and thus the solution $\mathcal{Q} = \tilde{S}R$ is

$$\mathcal{Q} = \left\{ \begin{array}{c} \rho_1 \\ 0 \\ \rho e_1 \end{array} \right\} \quad \text{for } x < -\tilde{a}t$$

$$\mathcal{Q} = \left\{ \begin{array}{c} \rho_1 + (\gamma-1)(\rho e_4 - \rho e_1)/2\tilde{a}^2 \\ (\gamma-1)(\rho e_1 - \rho e_4)/2\tilde{a} \\ \frac{1}{2}(\rho e_1 + \rho e_4) \end{array} \right\} \quad \text{for } -\tilde{a}t < x < 0$$

$$\mathcal{Q} = \left\{ \begin{array}{c} \rho_4 - (\gamma-1)(\rho e_4 - \rho e_1)/2\tilde{a}^2 \\ (\gamma-1)(\rho e_1 - \rho e_4)/2\tilde{a} \\ \frac{1}{2}(\rho e_1 + \rho e_4) \end{array} \right\} \quad \text{for } 0 < x < \tilde{a}t$$

$$\mathcal{Q} = \left\{ \begin{array}{c} \rho_4 \\ 0 \\ \rho e_4 \end{array} \right\} \quad \text{for } x > \tilde{a}t$$

The static temperature and pressure, velocity, and entropy are shown in Figs. 5.14 to 5.17 for the initial conditions $p_4/p_1 = 2$ and $T_4/T_1 = 1$ with $\gamma = 1.4$. The abscissa is x/t, normalized by a_1, whereby the solution at any time t can be obtained. The leftmost wave is a shock with exact speed $c_s/a_1 = -1.159479$. The Roe equations yield $c_s/a_1 = -\tilde{a}/a_1 = -1$. The center wave is a contact surface with exact velocity $c_c/a_1 = -0.247519$. The Roe equations yield $c_c/a_1 = 0$. The right wave is an expansion whose left and right boundaries are defined by the wave speeds $c_l/a_1 = 0.702978$ and $c_r/a_1 = 1$. The Roe equations yield a discontinuous expansion† (an *expansion shock*) with wave speed $c_s/a_1 = 1$. Overall, the solution of the Roe equations for the pressure, temperature, velocity, and entropy are an approximation of the exact solution of the Euler equations.

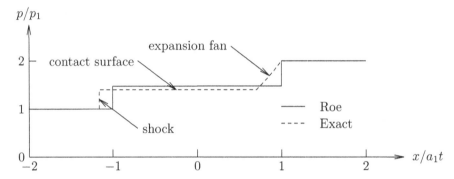

Fig. 5.14. Pressure for Riemann Shock Tube for $p_4/p_1 = 2$ and $T_4/T_1 = 1$

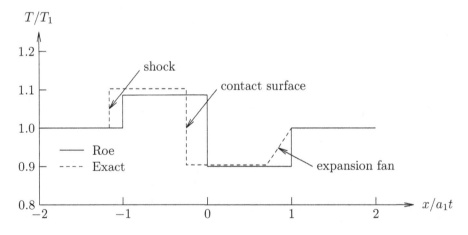

Fig. 5.15. Temperature for Riemann Shock Tube for $p_4/p_1 = 2$ and $T_4/T_1 = 1$

† This example is intended to illustrate the phenomenon of an expansion shock using Roe's method. A finite-volume algorithm solution of this example problem based on Roe's method would not necessarily yield an expansion shock due to the effects of spatial and temporal truncation errors. Detailed discussions of the concept of the *entropy condition* are presented in Hirsch (1988) and Laney (1998).

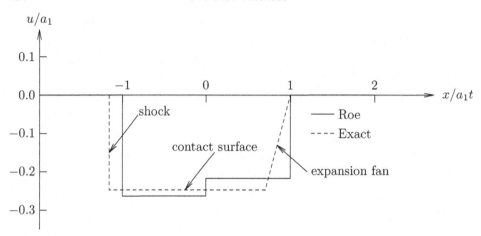

Fig. 5.16. Velocity for Riemann Shock Tube for $p_4/p_1 = 2$ and $T_4/T_1 = 1$

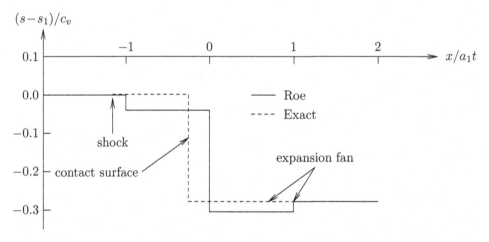

Fig. 5.17. Entropy for Riemann Shock Tube for $p_4/p_1 = 2$ and $T_4/T_1 = 1$

Harten (1983) proposed the following remedy to the expansion shock. The Roe eigenvalues $|\tilde{\lambda}_i|$ are replaced by the approximate eigenvalues $|\hat{\lambda}_i|$, where

$$|\hat{\lambda}_i| = \begin{cases} \tilde{\lambda}_i^2/4\varepsilon\hat{a} + \varepsilon\hat{a} & \text{for } |\tilde{\lambda}_i| < 2\varepsilon\hat{a} \\ |\tilde{\lambda}_i| & \text{for } |\tilde{\lambda}_i| \geq 2\varepsilon\hat{a} \end{cases} \tag{5.55}$$

where $\hat{a}$ is a suitable velocity scale (*e.g.*, $\hat{a} = \tilde{a}$) and ε is a small positive number (*e.g.*, $\varepsilon = 0.1$ when $\hat{a} = \tilde{a}$). Therefore, $|\hat{\lambda}_i| > 0$ for $i = 1, 2, 3$ and $\det(\tilde{S}|\tilde{\Lambda}|\tilde{S}^{-1}) > 0$ always.

5.4 Osher's Method

Osher developed an approximate Riemann solver† based on the concept of a series of path integrations in the space of solutions of the Euler equations (Engquist and Osher, 1980; Osher and Solomon, 1982; Osher and Chakravarthy, 1983).

5.4.1 Algorithm

Consider the Euler equations (3.9) in semi-discrete form:

$$\frac{dQ_i}{dt} + \frac{\left(F_{i+\frac{1}{2}} - F_{i-\frac{1}{2}}\right)}{\Delta x} = 0$$

where the flux $F_{i+\frac{1}{2}} = \mathcal{F}_{i+\frac{1}{2}}$ since the problem is one-dimensional. The Jacobian matrix $\mathcal{A}$ defined by

$$\mathcal{A} = \frac{\partial \mathcal{F}}{\partial \mathcal{Q}}$$

can be expressed as

$$\mathcal{A} = T\Lambda T^{-1}$$

where

$$\Lambda = \left\{ \begin{array}{ccc} \lambda_1 & 0 & 0 \\ 0 & \lambda_2 & 0 \\ 0 & 0 & \lambda_3 \end{array} \right\} \tag{5.56}$$

according to (2.15) and (2.16). The individual eigenvalues are split according to

$$\begin{array}{rcl} \lambda_k^+ & = & \max(\lambda_k, 0) \\ \lambda_k^- & = & \min(\lambda_k, 0) \end{array} \quad \text{for } k = 1, 2, 3 \tag{5.57}$$

Therefore,

$$\lambda_k = \lambda_k^+ + \lambda_k^- \quad \text{for } k = 1, 2, 3 \tag{5.58}$$

The diagonal matrix Λ is split according to

$$\Lambda = \Lambda^+ + \Lambda^- \tag{5.59}$$

---

† It could be also be argued that Osher's Method is a Flux Vector Split Method (Chapter 6) on the basis of (5.65). We choose to include it in this chapter on Godunov Methods since it is based on an exact solution of one of the five cases of the General Riemann Problem.

where

$$\Lambda^+ = \left\{ \begin{array}{ccc} \lambda_1^+ & 0 & 0 \\ 0 & \lambda_2^+ & 0 \\ 0 & 0 & \lambda_3^+ \end{array} \right\} \tag{5.60}$$

and

$$\Lambda^- = \left\{ \begin{array}{ccc} \lambda_1^- & 0 & 0 \\ 0 & \lambda_2^- & 0 \\ 0 & 0 & \lambda_3^- \end{array} \right\} \tag{5.61}$$

As an example, assume $0 < u < a$. Then

$$\Lambda^+ = \left\{ \begin{array}{ccc} u & 0 & 0 \\ 0 & u+a & 0 \\ 0 & 0 & 0 \end{array} \right\} \tag{5.62}$$

and

$$\Lambda^- = \left\{ \begin{array}{ccc} 0 & 0 & 0 \\ 0 & 0 & 0 \\ 0 & 0 & u-a \end{array} \right\} \tag{5.63}$$

The Jacobian matrix is therefore split according to

$$\mathcal{A} = \mathcal{A}^+ + \mathcal{A}^- \tag{5.64}$$

where

$$\begin{aligned} \mathcal{A}^+ &= T\Lambda^+ T^{-1} \\ \mathcal{A}^- &= T\Lambda^- T^{-1} \end{aligned}$$

It is straightforward[3] to show that λ_k^+ are eigenvalues of $\mathcal{A}^+$, λ_k^- are eigenvalues of $\mathcal{A}^-$, and the right eigenvector corresponding to both λ_k^+ and λ_k^- is r_k given by (2.13).

The flux is likewise split as

$$\mathcal{F} = \mathcal{F}^+ + \mathcal{F}^- \tag{5.65}$$

where

$$\frac{\partial \mathcal{F}^+}{\partial \mathcal{Q}} = \mathcal{A}^+ \tag{5.66}$$

$$\frac{\partial \mathcal{F}^-}{\partial \mathcal{Q}} = \mathcal{A}^- \tag{5.67}$$

and thus

$$F_{i+\frac{1}{2}} = \mathcal{F}^+_{i+\frac{1}{2}} + \mathcal{F}^-_{i+\frac{1}{2}} \tag{5.68}$$

The term $\mathcal{F}_{i+\frac{1}{2}}^{+}$ represents the contribution to the flux associated with the waves that move from left to right across the interface $x_{i+\frac{1}{2}}$ since $\mathcal{A}^{+} = T\Lambda^{+}T^{-1}$ and Λ^{+} is the diagonal matrix of eigenvalues that are positive at $x_{i+\frac{1}{2}}$. Similarly, the term $\mathcal{F}_{i+\frac{1}{2}}^{+}$ represents the contribution to the flux associated with the waves that move from right to left across the interface $x_{i+\frac{1}{2}}$ since $\mathcal{A}^{-} = T\Lambda^{-}T^{-1}$ and Λ^{-} is the diagonal matrix of eigenvalues that are negative at $x_{i+\frac{1}{2}}$. Thus, it is reasonable to use $Q_{i+\frac{1}{2}}^{l}$ to determine $\mathcal{F}_{i+\frac{1}{2}}^{+}$ and $Q_{i+\frac{1}{2}}^{r}$ to determine $\mathcal{F}_{i+\frac{1}{2}}^{-}$ according to

$$F_{i+\frac{1}{2}} = \mathcal{F}^{+}(Q_{i+\frac{1}{2}}^{l}) + \mathcal{F}^{-}(Q_{i+\frac{1}{2}}^{r}) \tag{5.69}$$

The essence of Osher's method is to determine an algorithm for $F_{i+\frac{1}{2}}$ that is equivalent to (5.69) but avoids entirely the need to actually compute either $\mathcal{F}^{+}$ or $\mathcal{F}^{-}$. Consider the integral

$$\int_{Q_l}^{Q_r} \mathcal{A}^{+} dQ$$

where we omit the subscript $i + \frac{1}{2}$ on Q^l and Q^r in the remainder of this section for simplicity. From (5.66),

$$\int_{Q_l}^{Q_r} \mathcal{A}^{+} dQ = \int_{Q_l}^{Q_r} \frac{\partial \mathcal{F}^{+}}{\partial Q} dQ = \mathcal{F}^{+}(Q^r) - \mathcal{F}^{+}(Q^l)$$

Similarly, from (5.67),

$$\int_{Q_l}^{Q_r} \mathcal{A}^{-} dQ = \int_{Q_l}^{Q_r} \frac{\partial \mathcal{F}^{-}}{\partial Q} dQ = \mathcal{F}^{-}(Q^r) - \mathcal{F}^{-}(Q^l)$$

and therefore

$$\mathcal{F}^{+}(Q^l) = \mathcal{F}^{+}(Q^r) - \int_{Q^l}^{Q^r} \mathcal{A}^{+} dQ$$

$$\mathcal{F}^{-}(Q^r) = \mathcal{F}^{-}(Q^l) + \int_{Q^l}^{Q^r} \mathcal{A}^{-} dQ$$

Thus,

$$\begin{aligned}
F_{i+\frac{1}{2}} &= \mathcal{F}^{+}(Q^l) + \mathcal{F}^{-}(Q^r) \\
&= \mathcal{F}^{+}(Q^l) + \underbrace{\mathcal{F}^{-}(Q^l) + \int_{Q^l}^{Q^r} \mathcal{A}^{-} dQ}_{\mathcal{F}^{-}(Q^r)} \\
&= \underbrace{\mathcal{F}^{+}(Q^l) + \mathcal{F}^{-}(Q^l)}_{\mathcal{F}(Q^l)} + \int_{Q^l}^{Q^r} \mathcal{A}^{-} dQ
\end{aligned}$$

Thus,

$$F_{i+\frac{1}{2}} = \mathcal{F}(Q^l) + \int_{Q^l}^{Q^r} \mathcal{A}^- dQ \tag{5.70}$$

which defines $F_{i+\frac{1}{2}}$ without specific mention of either $\mathcal{F}^+$ or $\mathcal{F}^-$. A similar expression for $F_{i+\frac{1}{2}}$ can be obtained in terms of $\mathcal{F}(Q^r)$ and the integral of $\mathcal{A}^+$.

Expression (5.70) is evaluated using the exact solution to the general Riemann problem for two expansions (Section 2.9.4). This expression is then assumed to be approximately valid for all cases. The wave structure is shown in Fig. 5.18. The leftmost state (denoted by $\mathcal{Q}_1$) corresponds to Q^l and the rightmost state (denoted by $\mathcal{Q}_4$) corresponds to Q^r.

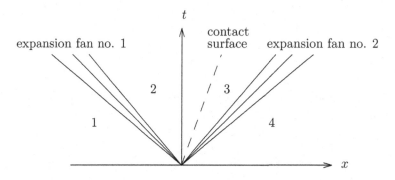

Fig. 5.18. General Riemann Problem: Case 4 (two expansions)

A closed-form expression for the entire solution can be obtained. The contact surface pressure p^* is (Exercise 5.7)

$$p^* = \left[\frac{a_1 + a_4 + \frac{(\gamma-1)}{2}(u_1 - u_4)}{a_1 p_1^{-\frac{(\gamma-1)}{2\gamma}} + a_4 p_4^{-\frac{(\gamma-1)}{2\gamma}}} \right]^{\frac{2\gamma}{\gamma-1}} \tag{5.71}$$

The solutions in Regions 2 and 3 are (Section 2.9.4)

$$
\begin{aligned}
c_c &= u_1 + \frac{2a_1}{(\gamma-1)}\left[1 - \left(\frac{p^*}{p_1}\right)^{\frac{(\gamma-1)}{2\gamma}}\right] \\
u_2 &= c_c \\
p_2 &= p^* \\
\rho_2 &= \rho_1 \left(\frac{p^*}{p_1}\right)^{\frac{1}{\gamma}} \\
u_3 &= c_c
\end{aligned}
$$

$$p_3 = p^*$$

$$p_3 = p_4 \left(\frac{p^*}{p_4} \right)^{\frac{1}{\gamma}}$$

The solutions within the left and right expansion fans are given by (2.146) to (2.148) and by (2.149) to (2.151), respectively. The boundaries of the expansion fans are given by (2.135) to (2.138).

We evaluate the integral in (5.70) by decomposing it into three separate components:

$$\int_{Q^l}^{Q^r} \mathcal{A}^- dQ = \underbrace{\int_1^2 \mathcal{A}^- dQ}_{\mathcal{I}_1} + \underbrace{\int_2^3 \mathcal{A}^- dQ}_{\mathcal{I}_2} + \underbrace{\int_3^4 \mathcal{A}^- dQ}_{\mathcal{I}_3} \qquad (5.72)$$

where $\mathcal{I}_1 = \int_1^2 \mathcal{A}^- dQ$ indicates the integral† from Region 1 to Region 2, and so on. For each integral $\mathcal{I}_j$ we consider an *integral curve* in the space of solutions Q defined by

$$\frac{dQ}{d\zeta} = \sigma_j(Q) \quad \text{for} \quad 0 \le \zeta \le \zeta_{1_j} \text{ on } \mathcal{I}_j$$

where σ_j depends on Q in some manner yet to be specified. Then

$$\begin{aligned}
\mathcal{I}_j &= \int_j^{j+1} \mathcal{A}^- dQ \\
&= \int_j^{j+1} \mathcal{A}^- \frac{dQ}{d\zeta} d\zeta \\
&= \int_j^{j+1} \mathcal{A}^- \sigma_j d\zeta \quad \text{for} \quad j = 1, 2, 3
\end{aligned}$$

A particularly useful choice for σ_j is one of the right eigenvectors (2.13) of $\mathcal{A}$:

$$\sigma_j = r_k$$

where r_k is one of the following:

$$r_1 = \left\{ \begin{array}{c} 1 \\ u \\ \frac{1}{2}u^2 \end{array} \right\}, \quad r_2 = \left\{ \begin{array}{c} 1 \\ u + a \\ H + ua \end{array} \right\}, \quad r_3 = \left\{ \begin{array}{c} 1 \\ u - a \\ H - ua \end{array} \right\}$$

and the value of k depends on the integral $\mathcal{I}_j$ in some manner yet to be

---

† Note that the actual integral is in the three-dimensional space of solutions Q, not in the $x-t$ plane.

specified. Note that $k \neq j$ in general. Since λ_k^- is an eigenvalue of $\mathcal{A}^-$ with a right eigenvector r_k,

$$\mathcal{A}^- r_k = \lambda_k^- r_k \quad \text{for} \quad k = 1, 2, 3$$

and

$$\mathcal{I}_j = \int_j^{j+1} \lambda_k^- r_k d\zeta$$

Assume furthermore that for each integral $\mathcal{I}_j$ there is a particular eigenvector r_k for which λ_k^- is *monotonic, i.e.,* λ_k^- changes sign at most one time on the integral curve

$$\frac{dQ}{d\zeta} = r_k \quad \text{for} \quad 0 \leq \zeta \leq \zeta_{1_j} \tag{5.73}$$

Let s denote the location in the space of solutions Q where λ_k changes sign if such a point exists. This is denoted the *sonic point.* Divide the integral $\mathcal{I}_j$ into two segments,

$$\mathcal{I}_j \ = \ \underbrace{\int_j^s \mathcal{A}^- dQ}_{\mathcal{I}_{j1}} + \underbrace{\int_s^{j+1} \mathcal{A}^- dQ}_{\mathcal{I}_{j2}}$$

$$= \ \int_j^s \lambda_k^- r_k d\zeta + \int_s^{j+1} \lambda_k^- r_k d\zeta \tag{5.74}$$

Assume that $\lambda_k < 0$ for $\mathcal{I}_{j1}$ and $\lambda_k \geq 0$ for $\mathcal{I}_{j2}$. Then $\lambda_k^- \neq 0$ for $\mathcal{I}_{j1}$ and $\lambda_k^- = 0$ for $\mathcal{I}_{j2}$. Hence

$$\begin{aligned}
\mathcal{I}_j \ &= \ \int_j^s \lambda_k^- r_k d\zeta \\
&= \ \int_j^s \mathcal{A}^- dQ \\
&= \ \int_j^s \left(\frac{\partial \mathcal{F}}{\partial Q} - \mathcal{A}^+ \right) dQ \\
&= \ \int_j^s \frac{\partial \mathcal{F}}{\partial Q} dQ - \int_j^s \mathcal{A}^+ dQ \\
&= \ \int_j^s \frac{\partial \mathcal{F}}{\partial Q} dQ - \int_j^s \lambda_k^+ r_k d\zeta \\
&= \ \int_j^s \frac{\partial \mathcal{F}}{\partial Q} dQ \quad \text{since } \lambda_k^+ = 0 \text{ when } \lambda_k^- \neq 0 \\
&= \ \mathcal{F}_s - \mathcal{F}_j
\end{aligned}$$

where $\mathcal{F}_s = \mathcal{F}(Q_s)$ and $\mathcal{F}_j = \mathcal{F}(Q_j)$ for $j = 1, \ldots, 4$. If $\lambda_k \geq 0$ for $\mathcal{I}_{j1}$ and

$\lambda_k < 0$ for $\mathcal{I}_{j2}$, then a similar analysis yields

$$\mathcal{I}_j = \mathcal{F}_{j+1} - \mathcal{F}_s$$

If $\lambda_k \geq 0$ for the entire path in $\mathcal{I}_j$, then

$$\mathcal{I}_j = 0$$

Similarly, if $\lambda_k < 0$ for the entire path in $\mathcal{I}_j$, then

$$\mathcal{I}_j = \mathcal{F}_{j+1} - \mathcal{F}_j$$

In summary,

$$\mathcal{I}_j = \begin{cases} \mathcal{F}_{j+1} - \mathcal{F}_j & \text{if } \lambda_k < 0 \text{ for } \mathcal{I}_j \\ \mathcal{F}_s - \mathcal{F}_j & \text{if } \lambda_k < 0 \text{ for } \mathcal{I}_{j1} \text{ and } \lambda_k \geq 0 \text{ for } \mathcal{I}_{j2} \\ \mathcal{F}_{j+1} - \mathcal{F}_s & \text{if } \lambda_k \geq 0 \text{ for } \mathcal{I}_{j1} \text{ and } \lambda_k < 0 \text{ for } \mathcal{I}_{j2} \\ 0 & \text{if } \lambda_k \geq 0 \text{ for } \mathcal{I}_j \end{cases} \qquad (5.75)$$

This form permits the integrals $\mathcal{I}_j$ to be expressed in terms of the flux $\mathcal{F}$ evaluated at specific points in the space of solutions $\mathcal{Q}$. Thus, $F_{i+\frac{1}{2}}$ in (5.69) is simply the sum of evaluations of $\mathcal{F}$ at these specific points.

It remains to demonstrate the two properties shown in Table 5.1 for each integral curve $\mathcal{I}_j$ and determine expressions for $\mathcal{Q}$ at the sonic points.

Table 5.1. *Requirements for Integral Curves*

Property	Requirement for $\mathcal{I}_j$
1	There exists a particular eigenvector r_k such that the integral curve $d\mathcal{Q}/d\zeta = r_k$ connects the solution points $\mathcal{Q}_j$ and $\mathcal{Q}_{j+1}$
2	The eigenvalue λ_k is monotonic on this integral curve

We proceed by construction for each integral beginning with $\mathcal{I}_1$. From (5.73) it is evident that ζ has the units of density. The density decreases monotonically from Region 1 to Region 2 since they are connected by an expansion. Thus, we may choose $\zeta = \rho - \rho_1$ with no loss of generality, and all flow variables are then functions of the single variable ρ in the expansion. Equation (5.73) then becomes

$$\frac{d\mathcal{Q}}{d\rho} = \begin{pmatrix} \dfrac{d\rho}{d\rho} \\[2mm] \dfrac{d\rho u}{d\rho} \\[2mm] \dfrac{d\rho e}{d\rho} \end{pmatrix}$$

The first element in $dQ/d\rho$ is 1. To determine the second element, we note that

$$u + \frac{2}{(\gamma-1)}a = u_1 + \frac{2}{(\gamma-1)}a_1 \qquad (5.76)$$

in the left expansion (Exercise 5.8). Thus,

$$\frac{du}{d\rho} = -\frac{2}{(\gamma-1)}\frac{da}{d\rho}$$

Moreover, the speed of sound in the expansion is

$$a = \sqrt{\frac{\gamma p}{\rho}} = \sqrt{\frac{\gamma p_1}{\rho_1^\gamma}}\,\rho^{(\gamma-1)/2}$$

and therefore

$$\frac{da}{d\rho} = \frac{(\gamma-1)}{2}\frac{a}{\rho}$$

Hence, the second term is

$$\begin{aligned}
\frac{d\rho u}{d\rho} &= u + \rho\frac{du}{d\rho} \\
&= u - \frac{2}{(\gamma-1)}\rho\frac{da}{d\rho} \\
&= u - a
\end{aligned}$$

and the third term is

$$\begin{aligned}
\frac{d\rho e}{d\rho} &= \frac{d}{d\rho}\left(\frac{1}{\gamma(\gamma-1)}\rho a^2 + \tfrac{1}{2}\rho u^2\right) \\
&= \frac{a^2}{(\gamma-1)} + \tfrac{1}{2}u^2 - ua \\
&= H - ua
\end{aligned}$$

where H is the total enthalpy. Thus,

$$\frac{dQ}{d\rho} = \begin{pmatrix} 1 \\ u - a \\ H - ua \end{pmatrix} = r_3 \qquad (5.77)$$

and Property 1 is satisfied. Next, we note that

$$\lambda_3 = u - a$$

Using (2.146) to (2.148),

$$u - a = \frac{x}{t} \quad \text{for} \quad c_{l_1} \le \frac{x}{t} \le c_{r_1}$$

Hence, λ_3 is monotonic on the integral curve

$$\frac{d\mathcal{Q}}{d\rho} = r_3$$

and Property 2 is satisfied. To evaluate (5.75), we consider the possibility of $\lambda_3 = 0$ on the integral curve. If this occurs, then the flow conditions at this sonic point are (Exercise 5.9)

$$
\begin{aligned}
u_{1_s} &= \frac{(\gamma-1)}{(\gamma+1)}u_1 + \frac{2}{(\gamma+1)}a_1 \\
a_{1_s} &= u_{1_s} \\
\rho_{1_s} &= \rho_1 \left(\frac{a_{1_s}}{a_1}\right)^{2/(\gamma-1)} \\
p_{1_s} &= p_1 \left(\frac{a_{1_s}}{a_1}\right)^{2\gamma/(\gamma-1)}
\end{aligned}
\tag{5.78}
$$

Since $a_{1_s} > 0$, the left sonic point exists only if

$$u_1 > -\frac{2}{(\gamma-1)}a_1 \tag{5.79}$$

The four possible cases for $\mathcal{I}_1$ are determined using (5.75) and are presented in Table 5.2.

For $\mathcal{I}_2$ we choose $\zeta = \rho - \rho_2$ and

$$\sigma_2 = r_1 = \left\{ \begin{array}{c} 1 \\ u \\ u^2 \end{array} \right\}$$

Then

$$
\begin{aligned}
\int_{\rho_2}^{\rho_3} r_1 d\rho &= r_1 \int_{\rho_2}^{\rho_3} d\rho \ \text{since}\ u_2 = u_3 \\
&= (\rho_3 - \rho_2)r_1 \\
&= \left(\begin{array}{c} \rho_3 - \rho_2 \\ \rho_3 u_3 - \rho_2 u_2 \\ \frac{1}{2}\rho_3 u_3^2 - \frac{1}{2}\rho_2 u_2^2 \end{array} \right) \\
&= \mathcal{Q}_3 - \mathcal{Q}_2
\end{aligned}
\tag{5.80}
$$

since

$$
\begin{aligned}
\rho_3 e_3 - \rho_2 e_2 &= \frac{p_3}{(\gamma-1)} + \frac{1}{2}\rho_3 u_3^2 - \frac{p_2}{(\gamma-1)} - \frac{1}{2}\rho_2 u_2^2 \\
&= \frac{1}{2}\rho_3 u_3^2 - \frac{1}{2}\rho_2 u_2^2
\end{aligned}
\tag{5.81}
$$

since $p_3 = p_2$. Thus, Property 1 is proven. Since $\lambda_1 = u$ is constant in Regions 2 and 3, λ_1 is monotonic† and hence Property 2 is proven. The two possible cases for $\mathcal{I}_2$ are presented in Table 5.2.

For $\mathcal{I}_3$ we choose $\zeta = \rho - \rho_3$ and consider

$$\frac{d\mathcal{Q}}{d\rho} = \begin{pmatrix} \dfrac{d\rho}{d\rho} \\[2mm] \dfrac{d\rho u}{d\rho} \\[2mm] \dfrac{d\rho e}{d\rho} \end{pmatrix}$$

The first element in $d\mathcal{Q}/d\rho$ is 1. To determine the second element, we note that

$$u - \frac{2}{(\gamma-1)}a = u_4 - \frac{2}{(\gamma-1)}a_4 \tag{5.82}$$

in the right expansion (Exercise 5.8). Thus,

$$\frac{du}{d\rho} = \frac{2}{(\gamma-1)}\frac{da}{d\rho}$$

The speed of sound in the right expansion is

$$a = \sqrt{\frac{\gamma p}{\rho}} = \sqrt{\frac{\gamma p_4}{\rho_4^\gamma}}\,\rho^{(\gamma-1)/2}$$

and therefore

$$\frac{da}{d\rho} = \frac{(\gamma-1)}{2}\frac{a}{\rho}$$

Hence, the second term is

$$\begin{aligned} \frac{d\rho u}{d\rho} &= u + \rho\frac{du}{d\rho} \\[1mm] &= u + \frac{2}{(\gamma-1)}\rho\frac{da}{d\rho} \\[1mm] &= u + a \end{aligned}$$

and the third term is

$$\frac{d\rho e}{d\rho} = \frac{d}{d\rho}\left(\frac{1}{\gamma(\gamma-1)}\rho a^2 + \tfrac{1}{2}u^2\right)$$

† If the integral curve vector σ_j is constant, the curve is denoted *linearly degenerate*.

$$
= \frac{a^2}{(\gamma-1)} + \tfrac{1}{2}u^2 + ua
$$
$$
= H + ua
$$

where H is the total enthalpy. Thus,

$$
\frac{dQ}{d\rho} = \begin{pmatrix} 1 \\ u+a \\ H+ua \end{pmatrix} = r_2 \tag{5.83}
$$

and Property 1 is satisfied. Next, we note that

$$
\lambda_2 = u + a
$$

Using (2.149) to (2.151),

$$
u + a = \frac{x}{t} \quad \text{for} \quad c_{l_2} \le \frac{x}{t} \le c_{r_2}
$$

Hence, λ_2 is monotonic on the integral curve

$$
\frac{dQ}{d\rho} = r_2
$$

and Property 2 is satisfied. A sonic point exists if $\lambda_2 = 0$ on the integral curve. If this occurs, then the flow conditions at this point are

$$
u_{4_s} = \frac{(\gamma-1)}{(\gamma+1)}u_4 - \frac{2}{(\gamma+1)}a_4
$$
$$
a_{4_s} = -u_{4_s}
$$
$$
\rho_{4_s} = \rho_4 \left(\frac{a_{4_s}}{a_4}\right)^{2/(\gamma-1)}
$$
$$
p_{4_s} = p_4 \left(\frac{a_{4_s}}{a_4}\right)^{2\gamma/(\gamma-1)} \tag{5.84}
$$

Since $a_{4_s} > 0$, the right sonic point exists only if

$$
u_4 < \frac{2}{(\gamma-1)}a_4 \tag{5.85}
$$

The four possible cases for $\mathcal{I}_3$ are determined using (5.75) and are presented in Table 5.2.

From Table 5.2, there are 32 different possible cases. However, four combinations of conditions for Regions 2 and 3 are inadmissible as indicated in Table 5.3. The admissible cases yield the flux formulas for $F_{i+\frac{1}{2}}$ shown in Table 5.4.

Table 5.2. *Evaluation of Integrals*

Integral	Case	$\int_j^{j+1} \mathcal{A}^- dQ$
$\mathcal{I}_1$	$u_1 - a_1 \geq 0, \ u_2 - a_2 \geq 0$	0
	$u_1 - a_1 \leq 0, \ u_2 - a_2 \leq 0$	$\mathcal{F}_2 - \mathcal{F}_1$
	$u_1 - a_1 \geq 0, \ u_2 - a_2 \leq 0$	$\mathcal{F}_2 - \mathcal{F}_{1_s}$
	$u_1 - a_1 \leq 0, \ u_2 - a_2 \geq 0$	$\mathcal{F}_{1_s} - \mathcal{F}_1$
$\mathcal{I}_2$	$c_c < 0$	$\mathcal{F}_3 - \mathcal{F}_2$
	$c_c \geq 0$	0
$\mathcal{I}_3$	$u_3 + a_3 \geq 0, \ u_4 + a_4 \geq 0$	0
	$u_3 + a_3 \leq 0, \ u_4 + a_4 \geq 0$	$\mathcal{F}_{4_s} - \mathcal{F}_3$
	$u_3 + a_3 \geq 0, \ u_4 + a_4 \leq 0$	$\mathcal{F}_4 - \mathcal{F}_{4_s}$
	$u_3 + a_3 \leq 0, \ u_4 + a_4 \leq 0$	$\mathcal{F}_4 - \mathcal{F}_3$

These integration paths employed for $\mathcal{I}_j$ are denoted the *physical ordering* or *P-ordering* (Toro, 1997). An alternate set of integration paths (the *original ordering* or *O-ordering*) was originally proposed by Osher; however, this approach is inaccurate or fails when $|u_1 - u_4| \gg 0$ (Toro, 1997).

Table 5.3. *Inadmissible Cases*

Case			Reason
$u_2 - a_2 \geq 0$	$c_c < 0$	$u_3 + a_3 \geq 0$	implies $a_2 < 0$
$u_2 - a_2 \geq 0$	$c_c < 0$	$u_3 + a_3 \leq 0$	implies $a_2 < 0$
$u_2 - a_2 \geq 0$	$c_c \geq 0$	$u_3 + a_3 \leq 0$	implies $a_3 < 0$
$u_2 - a_2 \leq 0$	$c_c \geq 0$	$u_3 + a_3 \leq 0$	implies $a_3 < 0$

Table 5.4. *Flux Formulas for Osher's Method*

	$\begin{aligned} u_1 - a_1 \geq 0 \\ u_4 + a_4 \geq 0 \end{aligned}$	$\begin{aligned} u_1 - a_1 \geq 0 \\ u_4 + a_4 \leq 0 \end{aligned}$	$\begin{aligned} u_1 - a_1 \leq 0 \\ u_4 + a_4 \geq 0 \end{aligned}$	$\begin{aligned} u_1 - a_1 \leq 0 \\ u_4 + a_4 \leq 0 \end{aligned}$
$a_2 \leq c_c$	$\mathcal{F}_1$	$\mathcal{F}_1 + \mathcal{F}_4 - \mathcal{F}_{4_s}$	$\mathcal{F}_{1_s}$	$\mathcal{F}_{1_s} + \mathcal{F}_4 - \mathcal{F}_{4_s}$
$0 \leq c_c \leq a_2$	$\mathcal{F}_1 + \mathcal{F}_2 - \mathcal{F}_{1_s}$	$\mathcal{F}_1 + \mathcal{F}_2 - \mathcal{F}_{1_s} + \mathcal{F}_4 - \mathcal{F}_{4_s}$	$\mathcal{F}_2$	$\mathcal{F}_2 + \mathcal{F}_4 - \mathcal{F}_{4_s}$
$-a_3 \leq c_c \leq 0$	$\mathcal{F}_1 - \mathcal{F}_{1_s} + \mathcal{F}_3$	$\mathcal{F}_1 - \mathcal{F}_{1_s} + \mathcal{F}_3 + \mathcal{F}_4 - \mathcal{F}_{4_s}$	$\mathcal{F}_3$	$\mathcal{F}_3 + \mathcal{F}_4 - \mathcal{F}_{4_s}$
$c_c \leq -a_3$	$\mathcal{F}_1 - \mathcal{F}_{1_s} + \mathcal{F}_{4_s}$	$\mathcal{F}_1 - \mathcal{F}_{1_s} + \mathcal{F}_4$	$\mathcal{F}_{4_s}$	$\mathcal{F}_4$

5.4.2 Stability

We now consider the semi-discrete stability of Osher's Method. The Euler equations (3.9) are

$$\frac{dQ_i}{dt} + \frac{\left(F_{i+\frac{1}{2}} - F_{i-\frac{1}{2}}\right)}{\Delta x} = 0$$

The flux $F_{i+\frac{1}{2}}$ depends on the left and right states:

$$F_{i+\frac{1}{2}} = F(Q^l_{i+\frac{1}{2}}, Q^r_{i+\frac{1}{2}})$$

We rewrite (3.9) as

$$\frac{dQ_i}{dt} = R_i(Q^l_{i+\frac{1}{2}}, Q^r_{i+\frac{1}{2}}, Q^l_{i-\frac{1}{2}}, Q^r_{i-\frac{1}{2}})$$

where

$$R_i = -\frac{\left(F_{i+\frac{1}{2}} - F_{i-\frac{1}{2}}\right)}{\Delta x}$$

We consider a specific case of Osher's Method from Table 5.4:

$$\left.\begin{array}{rcl} u_1 - a_1 & \geq & 0 \\ u_4 + a_4 & \geq & 0 \\ a_2 & \leq & c_c \end{array}\right\} \tag{5.86}$$

and assume that this condition holds at the cell interfaces $i \pm \frac{1}{2}$. Then

$$\begin{array}{rcl} F_{i+\frac{1}{2}} & = & \mathcal{F}(Q^l_{i+\frac{1}{2}}) \\ F_{i-\frac{1}{2}} & = & \mathcal{F}(Q^l_{i-\frac{1}{2}}) \end{array}$$

We further assume a simple first-order reconstruction:

$$\begin{array}{rcl} Q^l_{i+\frac{1}{2}} & = & Q_i \\ Q^l_{i-\frac{1}{2}} & = & Q_{i-1} \end{array}$$

Expanding in a Taylor series about Q_i and neglecting higher order terms,

$$\frac{dQ_i}{dt} = \frac{1}{\Delta x} A\left(Q_{i-1} - Q_i\right) \tag{5.87}$$

where $A = \mathcal{A}(Q_i)$ is treated as a constant.

For purposes of simplicity, we assume that the flow is periodic in x over a length $L = (M-1)\Delta x$, where $M = 2N+1$ and N is an integer. Consider

$Q(x,t)$ to be a continuous vector function that interpolates Q_i and expand $Q(x,t)$ in a Fourier series,

$$Q(x,t) = \sum_{l=-N+1}^{l=N} \hat{Q}_k(t)e^{\iota kx}$$

where $\iota = \sqrt{-1}$ and the wavenumber k depends on l according to

$$k = \frac{2\pi l}{L}$$

Substituting this into (5.87) yields

$$\frac{d\hat{Q}_k}{dt} = G\hat{Q}_k$$

where the amplification matrix G is

$$G = \frac{1}{\Delta x}\left(e^{-\iota k\Delta x} - 1\right)A$$

Since G is a constant multiple of A,

$$G = TDT^{-1}$$

where

$$D = \frac{1}{\Delta x}\left(e^{-\iota k\Delta x} - 1\right)\Lambda$$

It is evident that the eigenvalues of G are

$$\lambda_{G_m} = \frac{1}{\Delta x}\left(e^{-\iota k\Delta x} - 1\right)\lambda_m$$

where the λ_m are given in (2.12). Defining

$$\tilde{Q}_k = T^{-1}\hat{Q}_k$$

we have

$$\frac{d\tilde{Q}_k}{dt} = D\tilde{Q}_k$$

Defining

$$\tilde{Q}_k = \left\{ \begin{array}{c} \tilde{Q}_{k_1} \\ \tilde{Q}_{k_2} \\ \tilde{Q}_{k_3} \end{array} \right\}$$

the solution is then

$$\tilde{Q}_{k_m} = \tilde{Q}_{k_m}(0)e^{\lambda_{G_m}}$$

The condition for stability is therefore

$$\text{Real}\,(\lambda_{G_m}) \leq 0 \quad \text{for } m = 1, 2, 3$$

Now

$$\text{Real}\,(\lambda_{G_m}) = \frac{1}{\Delta x} \underbrace{(\cos k\Delta x - 1)}_{\leq 0} \lambda_m$$

From (5.86), $\lambda_m \geq 0$. Therefore, the algorithm is stable.[4]

5.4.3 Accuracy, Consistency, and Convergence

We now consider the problem described in Section 2.8. The initial condition is defined by (2.92) with $\epsilon = 0.1$ and the domain is $0 < \kappa x < 2\pi$. The first-order reconstruction (5.1) is employed. The norm, defined by (3.102), is evaluated at $\kappa a_o t = 7$, which is prior to the shock formation. The semi-discrete form (3.9) of the Euler equations is employed and the time integration is performed using a second-order Runge-Kutta method (Chapter 7).

The convergence is displayed in Fig. 5.19. The solution converges linearly[†] to the exact solution for sufficiently small Δt. The linear convergence is a direct consequence of the linear reconstruction (5.1). The computed result (using 100 cells) and exact solution are displayed[‡] in Fig. 5.20 for $\kappa a_o t = 7$ and $\mathcal{C} = 0.45$. The significant error in amplitude is attributable to the dissipative nature of the first-order reconstruction (5.1). The speed of the disturbance is accurately predicted, however.

The error in amplitude is significantly reduced by using a second-order reconstruction as indicated in Fig. 5.21, where results are shown using (4.20) with $\kappa = 0$ and no limiter together with results for the first-order reconstruction. The convergence for the second-order reconstruction (using second-order accurate integration in time) is shown in Fig. 5.22 and displays quadratic convergence.[§]

---

† The line in Fig. 5.19 is

$$||Q_i^{n,e} - Q_i^n||_{\kappa a_o \Delta t} = \left(\frac{\kappa a_o \Delta t}{5 \times 10^{-4}}\right) ||Q_i^{n,e} - Q_i^n||_{\kappa a_o \Delta t = 5 \times 10^{-4}}$$

‡ More precisely, the timestep $\kappa a_o \Delta t = 0.025$, which corresponds to $\mathcal{C} = 0.45$ at $t = 0$.
§ The line in Fig. 5.22 is

$$||Q_i^{n,e} - Q_i^n||_{\kappa a_o \Delta t} = \left(\frac{\kappa a_o \Delta t}{5 \times 10^{-4}}\right)^2 ||Q_i^{n,e} - Q_i^n||_{\kappa a_o \Delta t = 5 \times 10^{-4}}$$

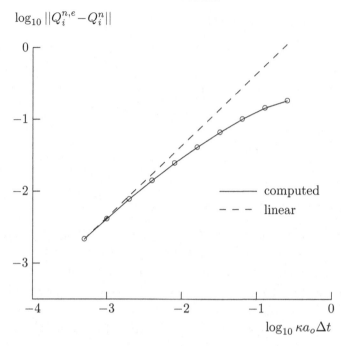

Fig. 5.19. Convergence for Osher's Method using a first-order reconstruction

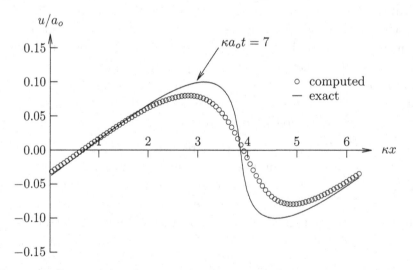

Fig. 5.20. Computed and exact solutions using a first-order reconstruction

Exercises

5.1   Prove the identities (5.12).

SOLUTION

a) $\Delta(f + g) = \Delta f + \Delta g$

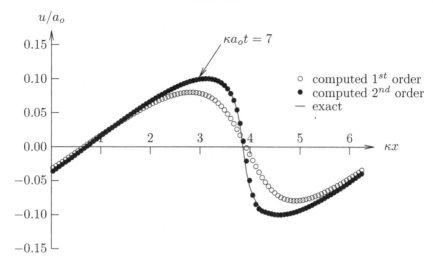

Fig. 5.21. Computed and exact solutions using first- and second-order reconstructions

The proof is straightforward.

b) $\Delta(fg) = \bar{f}\Delta g + \Delta f \bar{g}$

$$
\begin{aligned}
\bar{f}\Delta g + \Delta f \bar{g} &= \tfrac{1}{2}(f_l + f_r)(g_l - g_r) + (f_l - f_r)\tfrac{1}{2}(g_l + g_r) \\
&= \tfrac{1}{2}\left(f_l g_l - f_l g_r + f_r g_l - f_r g_r + f_l g_l + f_l g_r - f_r g_l - f_r g_r\right) \\
&= \tfrac{1}{2}\left(2f_l g_l - 2f_r g_r\right) \\
&= \Delta(fg)
\end{aligned}
$$

c) $\Delta(1/f) = -\Delta f / \hat{f}^2$

$$
\begin{aligned}
\Delta(1/f) &= f_l^{-1} - f_r^{-1} \\
&= (f_r - f_l)/f_l f_r \\
&= -\Delta f / \hat{f}^2
\end{aligned}
$$

5.2   Consider the quadratic function

$$
\begin{aligned}
f(\nu_1, \nu_2, \nu_3) = \; & a + b_1\nu_1 + b_2\nu_2 + b_3\nu_3 + c_1\nu_1^2 + c_2\nu_1\nu_2 + \\
& c_3\nu_1\nu_3 + c_4\nu_2^2 + c_5\nu_2\nu_3 + c_6\nu_3^2
\end{aligned}
$$

Prove that

$$
\Delta f = d_1\Delta\nu_1 + d_2\Delta\nu_2 + d_3\Delta\nu_3
$$

where d_1, d_2, and d_3 depend on the coefficients a, b_i, c_i and $\bar{\nu}_1, \bar{\nu}_2$ and $\bar{\nu}_3$.

5.3   Derive the intermediate matrices B and C used in the derivation of Roe's method.

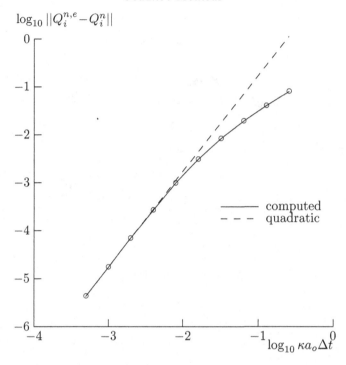

Fig. 5.22. Convergence for Osher's Method using a second-order reconstruction

SOLUTION

From (5.17) using (5.12),

$$\Delta \mathcal{Q} = \left\{ \begin{array}{c} 2\bar{\nu}_1 \Delta \nu_1 \\ \bar{\nu}_2 \Delta \nu_1 + \bar{\nu}_1 \Delta \nu_2 \\ \bar{\nu}_3 \Delta \nu_1/\gamma + \bar{\nu}_1 \Delta \nu_3/\gamma + (\gamma-1)\bar{\nu}_2 \Delta \nu_2/\gamma \end{array} \right\}$$

and therefore

$$B = \left\{ \begin{array}{ccc} 2\bar{\nu}_1 & 0 & 0 \\ \bar{\nu}_2 & \bar{\nu}_1 & 0 \\ \bar{\nu}_3/\gamma & (\gamma-1)\bar{\nu}_2/\gamma & \bar{\nu}_1/\gamma \end{array} \right\}$$

Also,

$$\Delta \mathcal{F} = \left\{ \begin{array}{c} \bar{\nu}_2 \Delta \nu_1 + \bar{\nu}_1 \Delta \nu_2 \\ (\gamma-1)\bar{\nu}_3 \Delta \nu_1/\gamma + (\gamma-1)\bar{\nu}_1 \Delta \nu_3/\gamma + (\gamma+1)\bar{\nu}_2 \Delta \nu_2/\gamma \\ \bar{\nu}_3 \Delta \nu_2 + \bar{\nu}_2 \Delta \nu_3 \end{array} \right\}$$

and therefore

$$C = \left\{ \begin{array}{ccc} \bar{\nu}_2 & \bar{\nu}_1 & 0 \\ (\gamma-1)\bar{\nu}_3/\gamma & (\gamma+1)\bar{\nu}_2/\gamma & (\gamma-1)\bar{\nu}_1/\gamma \\ 0 & \bar{\nu}_3 & \bar{\nu}_2 \end{array} \right\}$$

5.4   Derive the Roe matrix $\tilde{A}$ using $\tilde{A} = CB^{-1}$ and (5.20).

5.5   Derive the eigenvalues (5.23) and eigenvectors (5.24) of the Roe matrix $\tilde{A}$.

SOLUTION

By definition,

$$\det(\tilde{A} - \tilde{\lambda}I) = 0$$

Using (5.21),

$$\tilde{\lambda}\left\{\left[(3-\gamma)\tilde{u} - \tilde{\lambda}\right]\left[\gamma\tilde{u} - \tilde{\lambda}\right] - (\gamma-1)\left[\tilde{H} - (\gamma-1)\tilde{u}^2\right]\right\}$$
$$+ \left\{(\gamma-3)\frac{\tilde{u}^2}{2}(\gamma\tilde{u} - \tilde{\lambda}) - (\gamma-1)\left[(\gamma-1)\frac{\tilde{u}^3}{2} - \tilde{H}\tilde{u}\right]\right\} = 0$$

From (5.25),

$$\tilde{H} = \frac{\tilde{a}^2}{(\gamma-1)} + \tfrac{1}{2}\tilde{u}^2$$

which yields

$$-\tilde{\lambda}^3 + 3\tilde{u}\tilde{\lambda}^2 + (\tilde{a}^2 - 3\tilde{u}^2)\tilde{\lambda} + \tilde{u}^3 - \tilde{u}\tilde{a}^2 = 0$$

which directly factors into

$$(\tilde{\lambda} - \tilde{u})(\tilde{\lambda} - (\tilde{u} + \tilde{a}))(\tilde{\lambda} - (\tilde{u} - \tilde{a})) = 0$$

Denote the eigenvector $\tilde{e}_1$ corresponding to the eigenvalue $\tilde{\lambda}_1 = \tilde{u}$ as

$$\tilde{e}_1 = \left\{ \begin{array}{c} v_1 \\ v_2 \\ v_3 \end{array} \right\}$$

Then

$$\tilde{A}\tilde{e}_1 = \lambda_1 \tilde{e}_1$$

Thus,

$$v_2 = \tilde{u}v_1$$
$$\frac{(\gamma-3)}{2}\tilde{u}^2 v_1 + (3-\gamma)\tilde{u}v_2 + (\gamma-1)v_3 = \tilde{u}v_2$$
$$(-\tilde{H}\tilde{u} + \frac{(\gamma-1)}{2}\tilde{u}^3)v_1 + (\tilde{H} - (\gamma-1)\tilde{u})v_2 + \gamma\tilde{u}v_3 = \tilde{u}v_3$$

We may arbitrarily choose $v_1 = 1$. Using (5.25), the above equations yield $v_2 = \tilde{u}$ and $v_3 = \tfrac{1}{2}\tilde{u}^2$. The remaining eigenvectors are obtained using a similar approach.

5.6 Show that the equivalent Rankine-Hugoniot conditions (Section 2.3) for the Roe equations (5.11) are

$$\tilde{A}\Delta Q - u_w \Delta Q = 0$$

5.7 Derive the expression for the contact surface pressure (5.71) for the General Riemann Problem Case 4 (expansion-expansion).

SOLUTION

From (2.98) and (2.99), since $p^* < p_1$ and $p^* < p_4$ for Case 4,

$$\frac{2}{(\gamma-1)}\left[a_1\left(\frac{p^*}{p_1}\right)^{(\gamma-1)/2\gamma} + a_4\left(\frac{p^*}{p_4}\right)^{(\gamma-1)/2\gamma}\right] = u_1 - u_4 + \frac{2}{(\gamma-1)}(a_1 + a_4)$$

Solving for p^* yields

$$p^* = \left[\frac{a_1 + a_4 + \frac{(\gamma-1)}{2}(u_1 - u_4)}{a_1 p_1^{-\frac{(\gamma-1)}{2\gamma}} + a_4 p_4^{-\frac{(\gamma-1)}{2\gamma}}} \right]^{\frac{2\gamma}{\gamma-1}}$$

5.8   For the General Riemann Problem Case 4 (expansion-expansion), show that in the left expansion

$$u + \frac{2}{\gamma-1}a = u_1 + \frac{2}{\gamma-1}a_1$$

and in the right expansion

$$u - \frac{2}{\gamma-1}a = u_4 - \frac{2}{\gamma-1}a_4$$

5.9   Show that the left sonic point is defined by (5.78).

SOLUTION

At the left sonic point, $\lambda_{3_s} = u_{1_s} - a_{1_s} = 0$. Since the Riemann invariant (5.76) applies in the left expansion,

$$u_1 + \frac{2}{\gamma-1}a_1 = u_{1_s} + \frac{2}{\gamma-1}a_{1_s}$$

These equations may be solved to obtain

$$u_{1_s} = \frac{(\gamma-1)}{(\gamma+1)}u_1 + \frac{2}{(\gamma+1)}a_1$$

Since the flow is isentropic within the left expansion,

$$\rho_{1_s} = \rho_1 \left(\frac{a_{1_s}}{a_1} \right)^{2/(\gamma-1)}$$

$$p_{1_s} = p_1 \left(\frac{a_{1_s}}{a_1} \right)^{2\gamma/(\gamma-1)}$$

A similar analysis applied to the right expansion using the Riemann invariant (5.82) yields

$$u_4 - \frac{2}{\gamma-1}a_4 = u_{4_s} - \frac{2}{\gamma-1}a_{4_s}$$

and the right sonic point $\lambda_{2_s} = u_{2_s} + a_{2_s} = 0$ yields (5.84).

5.10   Derive the admissible cases in Table 5.4.

6

Flux Vector Splitting Methods

In fact, as far as characteristic modeling goes, aside from Riemann solvers, the only currently available alternative is flux vector splitting.

Culbert B. Laney (1998)

6.1 Introduction

Consider the Euler equations (3.9) in semi-discrete form:

$$\frac{dQ_i}{dt} + \frac{\left(F_{i+\frac{1}{2}} - F_{i-\frac{1}{2}}\right)}{\Delta x} = 0$$

where the flux $F_{i+\frac{1}{2}} = \mathcal{F}_{i+\frac{1}{2}}$ since the problem is one-dimensional. The basic concept of flux vector splitting is to decompose the flux vector F into two parts,

$$\mathcal{F} = \mathcal{F}^+ + \mathcal{F}^-$$

where

$$\frac{\partial \mathcal{F}^+}{\partial Q} \quad \text{has nonnegative eigenvalues}$$

$$\frac{\partial \mathcal{F}^-}{\partial Q} \quad \text{has nonpositive eigenvalues}$$

The term $\mathcal{F}^+$ represents the contribution to the flux associated with waves that move from left to right across the cell interface at $i + \frac{1}{2}$ since the eigenvalues of its Jacobian $\partial \mathcal{F}^+/\partial Q$ are positive (or zero). Thus it is reasonable to use $Q_{i+\frac{1}{2}}^l$ to evaluate $\mathcal{F}^+$. Similarly, the term $\mathcal{F}^-$ represents the contribution to the flux associated with waves that move from right to left across

147

the cell interface since the eigenvalues of its Jacobian $\partial \mathcal{F}^-/\partial Q$ are negative (or zero). Thus it is reasonable to use $Q^r_{i+\frac{1}{2}}$ to evaluate $\mathcal{F}^-$. Numerous algorithms for $\mathcal{F}^+$ and $\mathcal{F}^-$ have been developed. We present two methods in this chapter. For additional algorithms, see Laney (1998) and Toro (1997).

6.2 Steger and Warming's Method

Steger and Warming (1981) developed a flux vector split algorithm based on Euler's identity $\mathcal{F} = \mathcal{A}Q$; see also Sanders and Prendergast (1974).

6.2.1 Algorithm

From (2.15),

$$\mathcal{A} = T\Lambda T^{-1}$$

where

$$\Lambda = \left\{ \begin{array}{ccc} \lambda_1 & 0 & 0 \\ 0 & \lambda_2 & 0 \\ 0 & 0 & \lambda_3 \end{array} \right\}$$

with

$$\begin{array}{rcl} \lambda_1 & = & u \\ \lambda_2 & = & u + a \\ \lambda_3 & = & u - a \end{array}$$

and

$$T = \left\{ \begin{array}{ccc} 1 & 1 & 1 \\ u & u+a & u-a \\ \frac{1}{2}u^2 & H+ua & H-ua \end{array} \right\} \tag{6.1}$$

$$T^{-1} = \left\{ \begin{array}{ccc} 1 - \dfrac{(\gamma-1)}{2}\dfrac{u^2}{a^2} & (\gamma-1)\dfrac{u}{a^2} & -\dfrac{(\gamma-1)}{a^2} \\[3mm] \dfrac{(\gamma-1)}{4}\dfrac{u^2}{a^2} - \dfrac{1}{2}\dfrac{u}{a} & -\dfrac{(\gamma-1)}{2}\dfrac{u}{a^2} + \dfrac{1}{2a} & \dfrac{(\gamma-1)}{2a^2} \\[3mm] \dfrac{(\gamma-1)}{4}\dfrac{u^2}{a^2} + \dfrac{1}{2}\dfrac{u}{a} & -\dfrac{(\gamma-1)}{2}\dfrac{u}{a^2} - \dfrac{1}{2a} & \dfrac{(\gamma-1)}{2a^2} \end{array} \right\} \tag{6.2}$$

Steger and Warming assumed a *wave speed splitting*,

$$\lambda_i = \lambda_i^+ + \lambda_i^- \tag{6.3}$$

with

$$\begin{aligned}\lambda_i^+ &= \tfrac{1}{2}\left(\lambda_i + |\lambda_i|\right)\\\lambda_i^- &= \tfrac{1}{2}\left(\lambda_i - |\lambda_i|\right)\end{aligned} \tag{6.4}$$

This is equivalent to

$$\lambda_i^+ = \begin{cases} \lambda_i & \text{if } \lambda_i \geq 0 \\ 0 & \text{if } \lambda_i < 0 \end{cases} \tag{6.5}$$

$$\lambda_i^- = \begin{cases} 0 & \text{if } \lambda_i \geq 0 \\ \lambda_i & \text{if } \lambda_i < 0 \end{cases} \tag{6.6}$$

Thus

$$\Lambda = \Lambda^+ + \Lambda^-$$

where

$$\Lambda^+ = \left\{ \begin{matrix} \lambda_1^+ & 0 & 0 \\ 0 & \lambda_2^+ & 0 \\ 0 & 0 & \lambda_3^+ \end{matrix} \right\} \quad \text{and} \quad \Lambda^- = \left\{ \begin{matrix} \lambda_1^- & 0 & 0 \\ 0 & \lambda_2^- & 0 \\ 0 & 0 & \lambda_3^- \end{matrix} \right\} \tag{6.7}$$

Using (2.15) and (2.24),

$$\begin{aligned}\mathcal{F} &= \mathcal{A}\mathcal{Q}\\&= T\Lambda T^{-1}\mathcal{Q}\\&= \underbrace{T\Lambda^+ T^{-1}\mathcal{Q}}_{\mathcal{F}^+} + \underbrace{T\Lambda^- T^{-1}\mathcal{Q}}_{\mathcal{F}^-}\end{aligned} \tag{6.8}$$

The terms $\mathcal{F}^\pm$ can be written as (Exercise 6.1)

$$\mathcal{F}^\pm = \frac{(\gamma-1)}{\gamma}\rho\lambda_1^\pm \begin{bmatrix} 1 \\ u \\ \tfrac{1}{2}u^2 \end{bmatrix} + \frac{\rho}{2\gamma}\lambda_2^\pm \begin{bmatrix} 1 \\ u+a \\ H+ua \end{bmatrix} + \frac{\rho}{2\gamma}\lambda_3^\pm \begin{bmatrix} 1 \\ u-a \\ H-ua \end{bmatrix} \tag{6.9}$$

There are four possible cases for the eigenvalues $\lambda_i^\pm$ depending on the Mach number† M as indicated in Table 6.1 (Exercise 6.2). The fluxes $\mathcal{F}^+$ and $\mathcal{F}^-$ can be determined from (6.9) as described below.

For Case 1, all eigenvalues are negative, and the wave speed splitting

† The Mach number is strictly a nonnegative quantity defined by $M = |u|/a$. In the context of Steger and Warming's method, we define the Mach number $M = u/a$ and interpret a negative Mach number to imply $u < 0$.

Table 6.1. *Four Cases for Eigenvalues*

Case	Range	λ_1^+	λ_2^+	λ_3^+	λ_1^-	λ_2^-	λ_3^-
1	$M \leq -1$	0	0	0	λ_1	λ_2	λ_3
2	$-1 < M \leq 0$	0	λ_2	0	λ_1	0	λ_3
3	$0 < M \leq 1$	λ_1	λ_2	0	0	0	λ_3
4	$M \geq 1$	λ_1	λ_2	λ_3	0	0	0

yields $\lambda_i^+ = 0$ and $\lambda_i^- = \lambda_i < 0$ for $i = 1, 2, 3$. All waves intersecting the cell face at $x_{i+\frac{1}{2}}$ originate from the right as illustrated in Fig. 6.1, which is drawn for $u = -\frac{3}{2}a$. Thus, the flux is evaluated using the flow variables reconstructed to the right face:

$$\mathcal{F} = \frac{(\gamma-1)}{\gamma} \rho_r \lambda_{1_r} \begin{bmatrix} 1 \\ u_r \\ \frac{1}{2} u_r^2 \end{bmatrix} + \frac{\rho_r}{2\gamma} \lambda_{2_r} \begin{bmatrix} 1 \\ u_r + a_r \\ H_r + u_r a_r \end{bmatrix} + \frac{\rho_r}{2\gamma} \lambda_{3_r} \begin{bmatrix} 1 \\ u_r - a_r \\ H_r - u_r a_r \end{bmatrix} \tag{6.10}$$

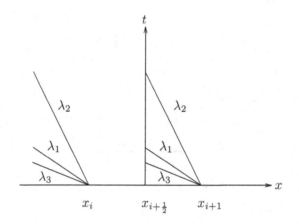

Fig. 6.1. Case 1

For Case 2, two eigenvalues are negative and one eigenvalue is positive. The wave speed splitting yields $\lambda_1^+ = 0$, $\lambda_2^+ = \lambda_2$, $\lambda_3^+ = 0$, $\lambda_1^- = \lambda_1$, $\lambda_2^- = 0$, and $\lambda_3^- = \lambda_3$. The wave corresponding to λ_2 intersects the cell face at $x_{i+\frac{1}{2}}$ from the left, while the waves corresponding to λ_1 and λ_3 intersect the cell face at $x_{i+\frac{1}{2}}$ from the right as illustrated in Fig. 6.2, which is drawn for $M = -\frac{1}{2}$. Thus, the flux is evaluated using the flow variables reconstructed to the right face and left face according to

$$\mathcal{F} = \frac{(\gamma-1)}{\gamma}\rho_r\lambda_{1_r}\begin{bmatrix} 1 \\ u_r \\ \frac{1}{2}u_r^2 \end{bmatrix} + \frac{\rho_l}{2\gamma}\lambda_{2_l}\begin{bmatrix} 1 \\ u_l + a_l \\ H_l + u_l a_l \end{bmatrix} + \frac{\rho_r}{2\gamma}\lambda_{3_r}\begin{bmatrix} 1 \\ u_r - a_r \\ H_r - u_r a_r \end{bmatrix}$$

$$(6.11)$$

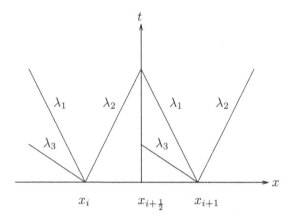

Fig. 6.2. Case 2

For Case 3, two eigenvalues are positive and one eigenvalue is negative. The wave speed splitting yields $\lambda_1^+ = \lambda_1$, $\lambda_2^+ = \lambda_2$, $\lambda_3^+ = 0$, $\lambda_1^- = 0$, $\lambda_2^- = 0$, and $\lambda_3^- = \lambda_3$. The waves corresponding to λ_1 and λ_2 intersect the cell face at $x_{i+\frac{1}{2}}$ from the left, while the wave corresponding to λ_3 intersects the cell face at $x_{i+\frac{1}{2}}$ from the right as illustrated in Fig. 6.3, which is drawn for $M = \frac{1}{2}$. Thus, the flux is evaluated using the flow variables reconstructed to the right face and left face according to

$$\mathcal{F} = \frac{(\gamma-1)}{\gamma}\rho_l\lambda_{1_l}\begin{bmatrix} 1 \\ u_l \\ \frac{1}{2}u_l^2 \end{bmatrix} + \frac{\rho_l}{2\gamma}\lambda_{2_l}\begin{bmatrix} 1 \\ u_l + a_l \\ H_l + u_l a_l \end{bmatrix} + \frac{\rho_r}{2\gamma}\lambda_{3_r}\begin{bmatrix} 1 \\ u_r - a_r \\ H_r - u_r a_r \end{bmatrix}$$

$$(6.12)$$

For Case 4, all eigenvalues are positive. The wave speed splitting yields $\lambda_i^+ = \lambda_i$ and $\lambda_i^- = 0$ for $i = 1, 2, 3$. All waves intersecting the cell face at $x_{i+\frac{1}{2}}$ originate from the left as illustrated in Fig. 6.3, which is drawn for $M = \frac{3}{2}$. Thus, the flux is evaluated using the flow variables reconstructed to the left face according to

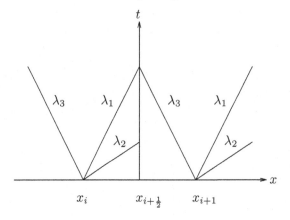

Fig. 6.3. Case 3

$$\mathcal{F} = \frac{(\gamma-1)}{\gamma}\rho_l\lambda_{1_l}\begin{bmatrix}1\\u_l\\\frac{1}{2}u_l^2\end{bmatrix} + \frac{\rho_l}{2\gamma}\lambda_{2_l}\begin{bmatrix}1\\u_l+a_l\\H_l+u_la_l\end{bmatrix} + \frac{\rho_l}{2\gamma}\lambda_{3_l}\begin{bmatrix}1\\u_l-a_l\\H_l-u_la_l\end{bmatrix}$$
(6.13)

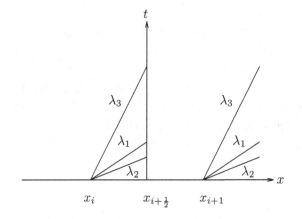

Fig. 6.4. Case 4

In practice, the value of M is the average Mach number at the interface, which may be defined as

$$M = \frac{u_l + u_r}{a_l + a_r}$$
(6.14)

The wave speeds $\lambda_i^\pm$ have a discontinuity in slope at $\lambda_i^\pm = 0$, which can lead to spurious numerical behavior at these locations. Therefore, $\lambda_i^{\pm}$ is replaced by (Laney, 1998)

$$\lambda_i^{\pm} = \tfrac{1}{2}\left(\lambda_i \pm \sqrt{\lambda_i^2 + \varepsilon^2 \hat{a}^2}\right) \tag{6.15}$$

where $\hat{a}$ is a suitable velocity (*e.g.*, $\hat{a} = a$) and ε is a small number (*e.g.*, $\varepsilon = 0.1$).

6.2.2 Stability

We consider the semi-discrete stability of Steger and Warming's Method. The Euler equations (3.9) are

$$\frac{dQ_i}{dt} + \frac{\left(F_{i+\frac{1}{2}} - F_{i-\frac{1}{2}}\right)}{\Delta x} = 0$$

We rewrite them as

$$\frac{dQ_i}{dt} = R_i(Q_{i+\frac{1}{2}}^l, Q_{i+\frac{1}{2}}^r, Q_{i-\frac{1}{2}}^l, Q_{i-\frac{1}{2}}^r)$$

where

$$R_i = -\frac{\left(F_{i+\frac{1}{2}} - F_{i-\frac{1}{2}}\right)}{\Delta x}$$

We assume a simple first-order reconstruction:

$$\begin{aligned}
Q_{i+\frac{1}{2}}^l &= Q_i \\
Q_{i+\frac{1}{2}}^r &= Q_{i+1}
\end{aligned}$$

Thus,

$$\begin{aligned}
F_{i+\frac{1}{2}} &= F_{i+\frac{1}{2}}^+(Q_{i+\frac{1}{2}}^l) + F_{i+\frac{1}{2}}^-(Q_{i+\frac{1}{2}}^r) \\
&= F^+(Q_i) + F^-(Q_{i+1}) \\
&= F^+(Q_i) + F^-(Q_i) + \frac{\partial F^-}{\partial Q}(Q_{i+1} - Q_i) + \mathcal{O}(\Delta Q^2)
\end{aligned}$$

and

$$\begin{aligned}
F_{i-\frac{1}{2}} &= F_{i-\frac{1}{2}}^+(Q_{i-\frac{1}{2}}^l) + F_{i-\frac{1}{2}}^-(Q_{i-\frac{1}{2}}^r) \\
&= F^+(Q_{i-1}) + F^-(Q_i) \\
&= F^+(Q_i) + \frac{\partial F^+}{\partial Q}(Q_{i-1} - Q_i) + F^-(Q_i) + \mathcal{O}(\Delta Q^2)
\end{aligned}$$

where we evaluate $\partial F^\pm/\partial Q$ at i and neglect its variation in x. Defining

$$
\begin{aligned}
A^+ &= \frac{\partial F^+}{\partial Q} \\
&= T\Lambda^+ T^{-1} \\
A^- &= \frac{\partial F^-}{\partial Q} \\
&= T\Lambda^- T^{-1}
\end{aligned}
$$

we then have

$$
\frac{dQ_i}{dt} = -\frac{1}{\Delta x}\left[A^- Q_{i+1} + (A^+ - A^-)Q_i - A^+ Q_{i-1}\right] \tag{6.16}
$$

For simplicity, we assume that the flow is periodic in x over a length $L = (M-1)\Delta x$,

$$
Q_1 = Q_M
$$

and we assume M is odd with $M = 2N + 1$. Consider $Q(x,t)$ to be a continuous vector function that interpolates Q_i,

$$
Q(x_i, t) = Q_i(t)
$$

Then the Fourier series (3.30) for $Q(x,t)$ is

$$
Q(x,t) = \sum_{l=-N+1}^{l=N} \hat{Q}_k(t)e^{\iota kx}
$$

where $\iota = \sqrt{-1}$ and the wavenumber k depends on l according to (3.31):

$$
k = \frac{2\pi l}{L}
$$

The Fourier coefficients $\hat{Q}_k(t)$ are complex vectors whose subscript k indicates an ordering with respect to the summation index l, *i.e.*, $\hat{Q}_k(t)$ indicates dependence on l (through (3.31)) and on t. Given the values of Q_i at some time t^n, the Fourier coefficients $\hat{Q}_k$ at t^n are obtained from (3.32).

Substituting the Fourier series into (6.16) yields

$$
\frac{d\hat{Q}_k}{dt} = G\hat{Q}_k \tag{6.17}
$$

where the amplification matrix G is

$$
G = -\frac{1}{\Delta x}\left[e^{\iota k\Delta x}A^- + A^+ - A^- - e^{-\iota k\Delta x}A^+\right]
$$

Now

$$G = TDT^{-1}$$

where D is the diagonal matrix

$$D = -\frac{1}{\Delta x}\left[e^{\iota k \Delta x}\Lambda^- + \Lambda^+ - \Lambda^- - e^{-\iota k \Delta x}\Lambda^+\right] \tag{6.18}$$

The eigenvalues of G are

$$\lambda_{G_m} = -\frac{1}{\Delta x}\left[e^{\iota k \Delta x}\lambda_m^- + \lambda_m^+ - \lambda_m^- - e^{-\iota k \Delta x}\lambda_m^+\right] \tag{6.19}$$

Defining

$$\tilde{Q}_k = T^{-1}\hat{Q}_k$$

gives

$$\frac{d\tilde{Q}_k}{dt} = D\tilde{Q}_k$$

Defining

$$\tilde{Q}_k = \left\{\begin{array}{c} \tilde{Q}_{k_1} \\ \tilde{Q}_{k_2} \\ \tilde{Q}_{k_3} \end{array}\right\}$$

then the solution is

$$\tilde{Q}_{k_m} = \tilde{Q}_{k_m}(0)e^{\lambda_{G_m}t} \quad \text{for} \quad m = 1, 2, 3$$

The condition for stability is therefore

$$\text{Real}(\lambda_{G_m}) \leq 0 \tag{6.20}$$

Now

$$\text{Real}(\lambda_{G_m}) = -\frac{1}{\Delta x}\underbrace{(\lambda_m^+ - \lambda_m^-)}_{\geq 0}\underbrace{(1 - \cos k\Delta x)}_{\geq 0}$$

and hence the stability condition (6.20) is satisfied.

6.2.3 Accuracy, Consistency, and Convergence

We consider the problem described in Section 2.8. The initial condition is defined by (2.92) with $\epsilon = 0.1$ and the domain is $0 < \kappa x < 2\pi$. The first-order reconstruction (5.1) is employed. The norm, defined by (3.102), is evaluated at $\kappa a_o t = 7$, which is prior to the shock formation. The semi-discrete form (3.9) of the Euler equations is employed and the time integration is performed using a second-order Runge-Kutta method (Chapter 7).

The convergence is displayed in Fig. 6.5. The solution converges linearly[†] to the exact solution for sufficiently small Δt. The linear convergence is a direct consequence of the linear reconstruction (5.1). The computed result (using 100 cells) and exact solution are displayed in Fig. 6.6 for $\kappa a_o t = 7$ and $C = 0.45$. The error in amplitude is attributable to the dissipative nature of the first-order reconstruction (5.1). The speed of the disturbance is accurately predicted, however.

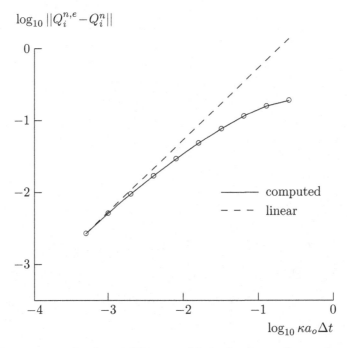

Fig. 6.5. Convergence for Steger-Warming Method using a first-order reconstruction

The error in amplitude is significantly reduced by using a second-order reconstruction as indicated in Fig. 6.8, where results are shown using (4.20) with $\kappa = 0$ and no limiter together with results for the first-order reconstruction. The convergence for the second-order reconstruction (using

---

[†] The line in Fig. 6.5 is

$$\|Q_i^{n,e} - Q_i^n\|_{\kappa a_o \Delta t} = \left(\frac{\kappa a_o \Delta t}{5 \times 10^{-4}}\right) \|Q_i^{n,e} - Q_i^n\|_{\kappa a_o \Delta t = 5 \times 10^{-4}}$$

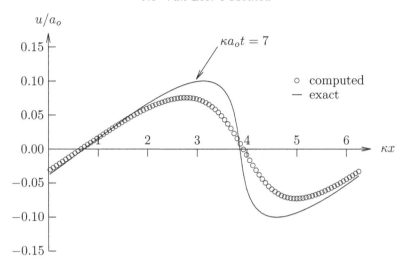

Fig. 6.6. Computed and exact solutions using a first-order reconstruction

second-order accurate integration in time) is shown in Fig. 6.7 and displays quadratic convergence.†

6.3 Van Leer's Method

Van Leer (1982) developed a flux vector split method based on the Mach number.

6.3.1 Algorithm

The flux vector $\mathcal{F}$ can be written (Exercise 6.4)

$$\mathcal{F} = \left\{ \begin{array}{c} \rho a M \\[2mm] \dfrac{\rho a^2}{\gamma}\left(\gamma M^2 + 1\right) \\[2mm] \rho a^3 M\left[\dfrac{1}{(\gamma-1)} + \tfrac{1}{2}M^2\right] \end{array} \right\} \tag{6.21}$$

† The line in Fig. 6.7 is

$$||Q_i^{n,e} - Q_i^n||_{\kappa a_o \Delta t} = \left(\frac{\kappa a_o \Delta t}{5 \times 10^{-4}}\right)^2 ||Q_i^{n,e} - Q_i^n||_{\kappa a_o \Delta t = 5 \times 10^{-4}}$$

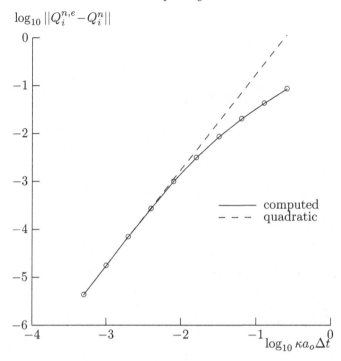

Fig. 6.7. Convergence for Steger-Warming Method using a second-order reconstruction

Each expression in (6.21) involves three quantities, namely, the density ρ, the speed of sound a, and the Mach number M. The term involving the Mach number is split into two parts, with ρ and a evaluated using Q^l or Q^r as appropriate. For the mass flux, the term involving the Mach number is simply M and is split according to

$$M = M^+ + M^- \tag{6.22}$$

The mass flux is taken to be

$$\rho u = \rho_l a_l M^+ + \rho_r a_r M^- \tag{6.23}$$

where the subscripts l and r imply that the quantities are evaluated using Q^l and Q^r, respectively. Van Leer proposed

$$M^+ = \begin{cases} 0 & \text{for } M \leq -1 \\ f_1^+ & \text{for } -1 \leq M \leq 1 \\ M & \text{for } M \geq 1 \end{cases} \tag{6.24}$$

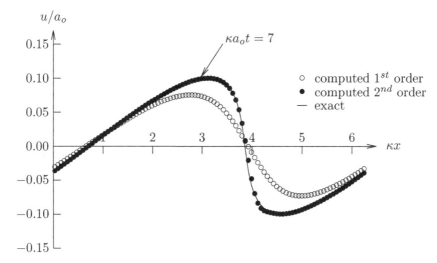

Fig. 6.8. Computed and exact solutions using first- and second-order reconstructions

and

$$M^- = \begin{cases} M & \text{for } M \leq -1 \\ f_1^- & \text{for } -1 \leq M \leq 1 \\ 0 & \text{for } M \geq 1 \end{cases} \qquad (6.25)$$

where M is the average Mach number at the interface, which may be defined as

$$M = \frac{u_l + u_r}{a_l + a_r} \qquad (6.26)$$

This yields

$$\rho u = \begin{cases} \rho_r a_r M & \text{for } M \leq -1 \\ \rho_l a_l f_1^+ + \rho_r a_r f_1^- & \text{for } -1 \leq M \leq 1 \\ \rho_l a_l M & \text{for } M \geq 1 \end{cases} \qquad (6.27)$$

For $M < -1$ the eigenvalues are negative, implying that all waves are moving to the left. It is therefore reasonable to use Q^r to compute ρ and a. Similarly, for $M > 1$ the eigenvalues are positive, implying that all waves are moving to the right. Hence, Q^l is employed to compute ρ and a.

It remains to determine the functions f_1^+ and f_1^-. They are chosen to satisfy (6.22) and to provide the continuity of $M^\pm$ and its first derivative with respect to M at $M = \pm 1$. It can be shown (Exercise 6.5) that

$$\begin{aligned} f_1^+ &= \tfrac{1}{4}(M+1)^2 \\ f_1^- &= -\tfrac{1}{4}(M-1)^2 \end{aligned} \qquad (6.28)$$

The complete expressions for $M^{\pm}$ are therefore

$$M^+ = \begin{cases} 0 & \text{for } M \leq -1 \\ \frac{1}{4}(M+1)^2 & \text{for } -1 \leq M \leq 1 \\ M & \text{for } M \geq 1 \end{cases} \tag{6.29}$$

and

$$M^- = \begin{cases} M & \text{for } M \leq -1 \\ -\frac{1}{4}(M-1)^2 & \text{for } -1 \leq M \leq 1 \\ 0 & \text{for } M \geq 1 \end{cases} \tag{6.30}$$

The functions are shown in Fig. 6.9. The mass flux is obtained from (6.23).

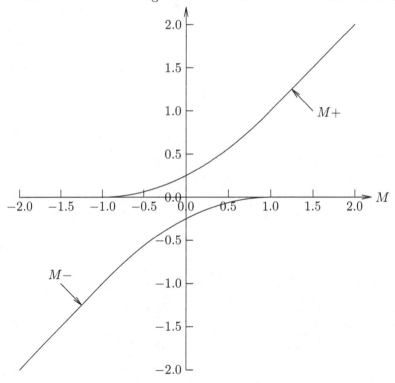

Fig. 6.9. M^+ and M^-

For the momentum flux, the term involving the Mach number is $(\gamma M^2 + 1)$ and is split according to

$$(\gamma M^2 + 1) = \left(\gamma M^2 + 1\right)^+ + \left(\gamma M^2 + 1\right)^- \tag{6.31}$$

The momentum flux is taken to be

$$\rho u^2 + p = \gamma^{-1} \rho_l a_l^2 \left(\gamma M^2 + 1\right)^+ + \gamma^{-1} \rho_r a_r^2 \left(\gamma M^2 + 1\right)^- \tag{6.32}$$

Van Leer proposed

$$
(\gamma M^2 + 1)^+ = \begin{cases} 0 & \text{for } M \leq 1 \\ f_2^+ & \text{for } -1 \leq M \leq 1 \\ \gamma M^2 + 1 & \text{for } M \geq 1 \end{cases} \tag{6.33}
$$

and

$$
(\gamma M^2 + 1)^- = \begin{cases} \gamma M^2 + 1 & \text{for } M \leq 1 \\ f_2^- & \text{for } -1 \leq M \leq 1 \\ 0 & \text{for } M \geq 1 \end{cases} \tag{6.34}
$$

This yields

$$
\rho u^2 + p = \begin{cases} \gamma^{-1}\rho_r a_r^2 (\gamma M^2 + 1) & \text{for } M \leq -1 \\ \gamma^{-1}\rho_r a_r^2 f_2^- + \gamma^{-1}\rho_l a_l^2 f_2^+ & \text{for } -1 \leq M \leq 1 \\ \gamma^{-1}\rho_l a_l^2 (\gamma M^2 + 1) & \text{for } M \geq 1 \end{cases} \tag{6.35}
$$

It remains to determine the functions f_2^+ and f_2^-. They are chosen to satisfy (6.31) and to provide the continuity of $(\gamma M^2 + 1)$ and its first derivative with respect to M at $M = \pm 1$. It can be shown (Exercise 6.6) that

$$
\begin{aligned}
f_2^+ &= \tfrac{1}{4}(M+1)^2\left[(\gamma-1)M+2\right] \\
f_2^- &= -\tfrac{1}{4}(M-1)^2\left[(\gamma-1)M-2\right]
\end{aligned} \tag{6.36}
$$

The complete expressions for $(\gamma M^2 + 1)^\pm$ are therefore

$$
(\gamma M^2 + 1)^+ = \begin{cases} 0 & \text{for } M \leq -1 \\ \tfrac{1}{4}(M+1)^2\left[(\gamma-1)M+2\right] & \text{for } -1 \leq M \leq 1 \\ \gamma M^2 + 1 & \text{for } M \geq 1 \end{cases} \tag{6.37}
$$

and

$$
(\gamma M^2 + 1)^- = \begin{cases} \gamma M^2 + 1 & \text{for } M \leq -1 \\ -\tfrac{1}{4}(M-1)^2\left[(\gamma-1)M-2\right] & \text{for } -1 \leq M \leq 1 \\ 0 & \text{for } M \geq 1 \end{cases} \tag{6.38}
$$

The functions are shown in Fig. 6.10. The momentum flux is obtained from (6.32).

For the energy flux, the term involving the Mach number is $M[(\gamma-1)^{-1} + \tfrac{1}{2}M^2]$ and is split according to

$$
M[(\gamma-1)^{-1}+\tfrac{1}{2}M^2] = M[(\gamma-1)^{-1}+\tfrac{1}{2}M^2]^+ + M[(\gamma-1)^{-1}+\tfrac{1}{2}M^2]^- \tag{6.39}
$$

The energy flux is taken to be

$$
(\rho e + p)u = \rho_l a_l^3 M[(\gamma-1)^{-1}+\tfrac{1}{2}M^2]^+ + \rho_r a_r^3 M[(\gamma-1)^{-1}+\tfrac{1}{2}M^2]^- \tag{6.40}
$$

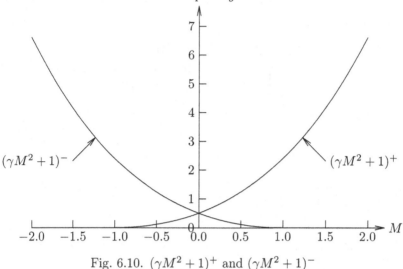

Fig. 6.10. $(\gamma M^2 + 1)^+$ and $(\gamma M^2 + 1)^-$

Van Leer proposed

$$M[(\gamma - 1)^{-1} + \tfrac{1}{2}M^2]^+ = \begin{cases} 0 & \text{for } M \leq -1 \\ f_3^+ & \text{for } -1 \leq M \leq 1 \\ M[(\gamma - 1)^{-1} + \tfrac{1}{2}M^2] & \text{for } M \geq 1 \end{cases}$$

(6.41)

and

$$M[(\gamma - 1)^{-1} + \tfrac{1}{2}M^2]^- = \begin{cases} M[(\gamma - 1)^{-1} + \tfrac{1}{2}M^2] & \text{for } M \leq -1 \\ f_3^- & \text{for } -1 \leq M \leq 1 \\ 0 & \text{for } M \geq 1 \end{cases}$$

(6.42)

This yields

$$(\rho e + p)u = \begin{cases} \rho_r a_r^3 M[(\gamma - 1)^{-1} + \tfrac{1}{2}M^2] & \text{for } M \leq -1 \\ \rho_l a_l^3 f_3^+ + \rho_r a_r^3 f_3^- & \text{for } -1 \leq M \leq 1 \\ \rho_l a_l^3 M[(\gamma - 1)^{-1} + \tfrac{1}{2}M^2] & \text{for } M \geq 1 \end{cases}$$

(6.43)

It remains to determine the functions f_3^+ and f_3^-. They are chosen to satisfy (6.39) and to provide the continuity of $M[(\gamma-1)^{-1}+\tfrac{1}{2}M^2]$ and its derivative with respect to M at $M = \pm 1$. It can be shown (Exercise 6.7) that

$$f_3^+ = \tfrac{1}{8}(\gamma + 1)^{-1}(\gamma - 1)^{-1}(M + 1)^2 \left[(\gamma - 1)M + 2\right]^2$$
$$f_3^- = -\tfrac{1}{8}(\gamma + 1)^{-1}(\gamma - 1)^{-1}(M - 1)^2 \left[(\gamma - 1)M - 2\right]^2$$

(6.44)

The complete expressions for $M[(\gamma - 1)^{-1} + \tfrac{1}{2}M^2]^\pm$ are therefore

$$M[(\gamma - 1)^{-1} + \tfrac{1}{2}M^2]^+ =$$

$$
\begin{cases}
0 & \text{for } M \le -1 \\
\frac{1}{8}(\gamma + 1)^{-1}(\gamma - 1)^{-1}(M + 1)^2 \left[(\gamma - 1)M + 2\right]^2 & \text{for } -1 \le M \le 1 \quad (6.45)\\
M[(\gamma - 1)^{-1} + \frac{1}{2}M^2] & \text{for } M \ge 1
\end{cases}
$$

and

$$
M[(\gamma - 1)^{-1} + \tfrac{1}{2}M^2]^- =
$$
$$
\begin{cases}
M[(\gamma - 1)^{-1} + \frac{1}{2}M^2] & \text{for } M \le -1 \\
-\frac{1}{8}(\gamma + 1)^{-1}(\gamma - 1)^{-1}(M - 1)^2 \left[(\gamma - 1)M - 2\right]^2 & \text{for } -1 \le M \le 1 \quad (6.46)\\
0 & \text{for } M \ge 1
\end{cases}
$$

The functions are shown in Fig. 6.11. The energy flux is obtained from (6.40).

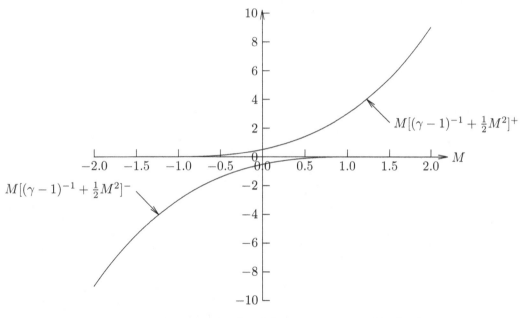

Fig. 6.11. $M[(\gamma - 1)^{-1} + \frac{1}{2}M^2]^+$ and $M[(\gamma - 1)^{-1} + \frac{1}{2}M^2]^-$

6.3.2 Stability

The stability analysis for Van Leer's method is identical to that in Section 6.2.2 provided that the fluxes F^+ and F^- satisfy

$$
\frac{\partial F^+}{\partial Q} = T\hat{\Lambda}^+ T^{-1} \tag{6.47}
$$

$$
\frac{\partial F^-}{\partial Q} = T\hat{\Lambda}^- T^{-1} \tag{6.48}
$$

where $\hat{\Lambda}^+$ and $\hat{\Lambda}^-$ are diagonal matrices whose elements are the eigenvalues $\hat{\lambda}_i^+$ of $\partial F^+/\partial Q$ and $\hat{\lambda}_i^-$ of $\partial F^-/\partial Q$, respectively, and

$$\hat{\lambda}_i^+ \quad \text{are nonnegative} \tag{6.49}$$

$$\hat{\lambda}_i^- \quad \text{are nonpositive} \tag{6.50}$$

We now consider (6.49). There are three possible cases: For $M \geq 1$, $F^+ = F$ and the eigenvalues of $\partial F^+/\partial Q$ are given by (2.12). Since $M \geq 1$, all eigenvalues are nonnegative. For $M \leq 1$, $F^+ = 0$ and there is no contribution to the stability analysis. For $-1 < M < 1$, the eigenvalues were determined by Van Leer (1982). Since the energy flux may be expressed as

$$\rho a^3 f_3^+ = \frac{1}{2(\gamma^2-1)}\rho a^3 \frac{(f_2^+)^2}{f_1^+} \quad \text{for } -1 < M < 1$$

the third row of $\partial F^+/\partial Q$ is a linear combination of the first and second rows, and hence $\det(\partial F^+/\partial Q) = 0$ for $-1 < M < 1$. Thus, $\hat{\lambda}_1 = 0$ is an eigenvalue of $\det(\partial F^+/\partial Q)$ for $-1 < M < 1$. The remaining two eigenvalues are solutions of the quadratic equation (Van Leer, 1982)

$$\hat{\lambda}^2 - \tfrac{3}{2}a(M+1)\left\{1 - \frac{(\gamma-1)(M-1)}{12\gamma(\gamma+1)}\left[\gamma(M-1)^2 + 2\gamma(M-1) - 2(\gamma+3)\right]\right\}\hat{\lambda}$$
$$+ \tfrac{1}{4}a^2(M+1)^3\left\{1 - \frac{(M-1)}{8\gamma(\gamma+1)}\left[4\gamma(\gamma-1)(M-1) + (\gamma+1)(3-\gamma)\right]\right\} = 0 \tag{6.51}$$

Van Leer concluded that both roots of this equation are positive provided $1 < \gamma < 3$. A similar analysis proves (6.50).

6.3.3 Accuracy, Consistency, and Convergence

We consider the problem described in Section 2.8. The initial condition is defined by (2.92) with $\epsilon = 0.1$ and the domain is $0 < \kappa x < 2\pi$. The first-order reconstruction in (5.1) is employed. The norm, defined by (3.102), is evaluated at $\kappa a_o t = 7$, which is prior to the shock formation. The semi-discrete form (3.9) of the Euler equations is employed and the time integration is performed using a second-order Runge-Kutta method (Chapter 7).

The convergence is displayed in Fig. 6.12. The solution converges linearly[†]

† The line in Fig. 6.12 is

$$\|Q_i^{n,e} - Q_i^n\|_{\kappa a_o \Delta t} = \left(\frac{\kappa a_o \Delta t}{5 \times 10^{-4}}\right)\|Q_i^{n,e} - Q_i^n\|_{\kappa a_o \Delta t = 5 \times 10^{-4}}$$

to the exact solution for sufficiently small Δt. The linear convergence is a direct consequence of the linear reconstruction in (5.1). The computed result (using 100 cells) and exact solution are displayed in Fig. 6.13 for $\kappa a_o t = 7$ and $\mathcal{C} = 0.45$. The error in amplitude is attributable to the dissipative nature of the first-order reconstruction in (5.1). The speed of the disturbance is accurately predicted, however.

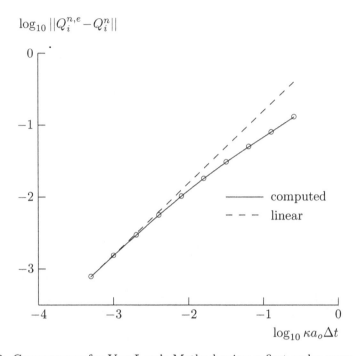

Fig. 6.12. Convergence for Van Leer's Method using a first-order reconstruction

The error in amplitude is significantly reduced by using a second-order reconstruction as indicated in Fig. 6.14, where the results are shown using (4.20) with $\kappa = 0$ and no limiter together with results for the first-order reconstruction. The convergence for the second-order reconstruction (using second-order accurate integration in time) is shown in Fig. 6.15 and displays quadratic convergence.†

---

† The line in Fig. 6.15 is

$$||Q_i^{n,e} - Q_i^n||_{\kappa a_o \Delta t} = \left(\frac{\kappa a_o \Delta t}{5 \times 10^{-4}} \right)^2 ||Q_i^{n,e} - Q_i^n||_{\kappa a_o \Delta t = 5 \times 10^{-4}}$$

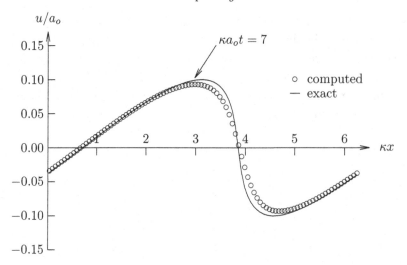

Fig. 6.13. Computed and exact solutions using a first-order reconstruction

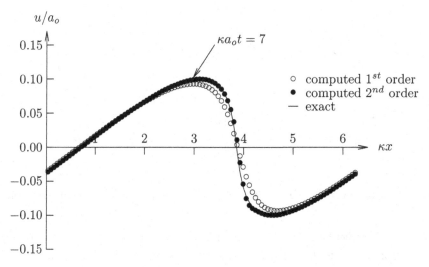

Fig. 6.14. Computed and exact solutions using first- and second-order reconstructions

Exercises

6.1   Derive the expression (6.9).

SOLUTION

From (6.8),

$$\mathcal{F}^{\pm} = T\Lambda^{\pm}T^{-1}\mathcal{Q}$$

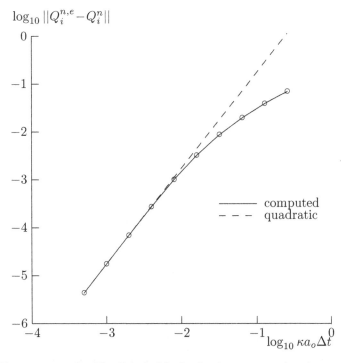

Fig. 6.15. Convergence for Van Leer's Method using a second-order reconstruction

Using (6.2),

$$T^{-1}\mathcal{Q} = \left\{ \begin{array}{c} (\gamma-1)\rho/\gamma \\ \rho/2\gamma \\ \rho/2\gamma \end{array} \right\}$$

Using (6.1), the expression for $\mathcal{F}^\pm$ in (6.9) is obtained.

6.2 Derive Table 6.1 using (2.12), (6.5), and (6.6).

6.3 Determine the discontinuity in the slope of $\lambda_i^\pm$ at $\lambda_i^\pm = 0$. Show that (6.15) provides a continuous slope at $\lambda_i^\pm = 0$.

SOLUTION

From (6.4),

$$\frac{d\lambda_i^+}{d\lambda_i} = \frac{d}{d\lambda_i}\tfrac{1}{2}\left(\lambda_i + |\lambda_i|\right) = \left\{ \begin{array}{ll} 1 & \lambda_i > 0 \\ 0 & \lambda_i < 0 \end{array} \right.$$

$$\frac{d\lambda_i^-}{d\lambda_i} = \frac{d}{d\lambda_i}\tfrac{1}{2}\left(\lambda_i - |\lambda_i|\right) = \left\{ \begin{array}{ll} 0 & \lambda_i > 0 \\ 1 & \lambda_i < 0 \end{array} \right.$$

From (6.15),

$$\frac{d\lambda_i^\pm}{d\lambda_i} = \tfrac{1}{2}\left(1 \pm \frac{\lambda_i}{\sqrt{\lambda_i^2 + \varepsilon^2 \hat{a}^2}}\right)$$

and thus

$$\frac{d\lambda_i^{\pm}}{d\lambda_i}\bigg|_{\lambda_i \to 0} = \tfrac{1}{2}$$

6.4   Derive the expression (6.21).

6.5   Show that the expressions (6.28) for $M^{\pm}$ satisfy (6.22), are contin-
        uous, and have continuous first derivatives with respect to M at
        $M = \pm 1$.

SOLUTION

From (6.28),

$$\begin{aligned}
f_1^+ + f_1^- &= \tfrac{1}{4}(M+1)^2 - \tfrac{1}{4}(M-1)^2 \\
&= \tfrac{1}{4}\left(M^2 + 2M + 1 - M^2 + 2M - 1\right) \\
&= M
\end{aligned}$$

and therefore (6.22) is satisfied. By inspection, the expressions for M^+ in (6.29) and for M^- in (6.30) are continuous at $M = \pm 1$. From (6.29),

$$\begin{aligned}
\frac{dM^+}{dM} &= \tfrac{1}{2}(M+1) \\
&= 0 \quad \text{at } M = -1 \\
&= 1 \quad \text{at } M = 1
\end{aligned}$$

and thus dM^+/dM is continuous at $M = \pm 1$. A similar results holds for M^-.

6.6   Show that the expressions (6.36) for $(\gamma M^2 + 1)^{\pm}$ satisfy (6.31), are
        continuous, and have continuous first derivatives with respect to M
        at $M = \pm 1$.

6.7   Show that the expressions (6.44) for $M[(\gamma - 1)^{-1} + \tfrac{1}{2}M^2]^{\pm}$ satisfy
        (6.39), are continuous, and have continuous first derivatives with
        respect to M at $M = \pm 1$.

SOLUTION

From (6.44), the identity (6.39) is obtained by substitution. By inspection, the expres-
sions for $M[(\gamma - 1)^{-1} + \tfrac{1}{2}M^2]^+$ in (6.45) and for $M[(\gamma - 1)^{-1} + \tfrac{1}{2}M^2]^-$ in (6.46) are
continuous at $M = \pm 1$. The derivatives of $M[(\gamma - 1)^{-1} + \tfrac{1}{2}M^2]^{\pm}$ can be shown to be
continuous at $M = \pm 1$ by straightforward differentiation.

6.8   Prove that the momentum flux splitting in (6.32) implies a pressure
        splitting (Laney, 1998):

$$p^+ = p \begin{cases} 0 & M \leq 1 \\ \left(\frac{M+1}{2}\right)^2 (2 - M) & -1 < M < 1 \\ 1 & M \geq 1 \end{cases}$$

$$p^- = p \begin{cases} 1 & M \leq 1 \\ -\left(\frac{M-1}{2}\right)^2 (2 + M) & -1 < M < 1 \\ 0 & M \geq 1 \end{cases}$$

6.9 Derivation of the eigenvalues of F^+ for Van Leer's method requires, for example, an expression for $\partial(\rho a M^+)/\partial\rho$. Find this expression in terms of a, M and γ.

SOLUTION

By definition, for $-1 < M < 1$,

$$\frac{\partial}{\partial\rho}\left(\rho a M^+\right) = \frac{\partial}{\partial\rho}\left[\rho a \left(\frac{M+1}{2}\right)^2\right]$$

where $\partial/\partial\rho$ implies holding ρu and ρe constant. Thus,

$$\frac{\partial}{\partial\rho}\left(\rho a M^+\right) = a\left(\frac{M+1}{2}\right)^2 + \rho\left(\frac{M+1}{2}\right)^2\frac{\partial a}{\partial\rho} + \rho a \left(\frac{M+1}{2}\right)\frac{\partial M}{\partial\rho}$$

Now

$$\rho e = \frac{1}{\gamma(\gamma-1)}\rho a^2 + \frac{(\rho u)^2}{\rho}$$

and thus differentiating with respect to ρ gives

$$\frac{\partial a}{\partial\rho} = \frac{a}{2\rho}\left[\gamma(\gamma-1)M^2 - 1\right]$$

Now

$$\rho a^2 M^2 = \frac{(\rho u)^2}{\rho}$$

and thus differentiating with respect to ρ gives

$$M^2\frac{\partial\rho a^2}{\partial\rho} + 2\rho a^2 M\frac{\partial M}{\partial\rho} = -\left(\frac{\rho u}{\rho}\right)^2$$

Inverting the definition of ρe gives

$$\rho a^2 = \gamma(\gamma-1)\left[\rho e - \frac{(\rho u)^2}{\rho}\right]$$

and thus

$$\frac{\partial\rho a^2}{\partial\rho} = \gamma(\gamma-1)u^2$$

Thus,

$$\frac{\partial M}{\partial\rho} = -\frac{M}{2\rho}\left[\gamma(\gamma-1)M^2 + 1\right]$$

Substituting into the initial equation gives

$$\frac{\partial}{\partial\rho}\left(\rho a M^+\right) = -\frac{a}{8}\left(M^2 - 1\right)\left[\gamma(\gamma-1)M^2 + 1\right]$$

6.10 Show that the eigenvalues $\hat{\lambda}^+$ obtained from (6.51) are positive for $1 < \gamma < 3$.

7

Temporal Quadrature

A great many if not all the problems in mathematics may be so formulated that they consist in finding from given data the values of certain unknown quantities subject to certain conditions.

Carl Runge (1912)

7.1 Introduction

We consider algorithms for the temporal quadrature of the semi-discrete Euler equations

$$\frac{dQ_i}{dt} + \frac{\left(F_{i+\frac{1}{2}} - F_{i-\frac{1}{2}}\right)}{\Delta x} = 0 \qquad (7.1)$$

Temporal quadrature algorithms can be categorized as *explicit* or *implicit*. For explicit methods, the values Q_i at each timestep (or subiterate within the timestep) may be evaluated independently. This is the simplest approach. However, explicit methods are restricted by the Courant-Friedrichs-Lewy condition introduced in Section 3.9.2. Explicit methods are easily parallelized.† For implicit methods, all values of Q_i at each timestep must be solved simultaneously. Implicit methods typically are not subject to the Courant-Friedrichs-Lewy condition. These methods can also be extended to parallel computation, albeit with greater effort.

---

† Parallelization is the use of multiple central processing units (cpus) to solve a single problem. Recent results in the development of parallel algorithms for computational fluid dynamics are presented in Matsuno (2003). A history of parallel computers (through the late 1980s) is presented in Hockney and Jessope (1988).

We rewrite the semi-discrete Euler equations as

$$\frac{dQ_i}{dt} = R_i \qquad (7.2)$$

with

$$R_i = -\frac{\left(F_{i+\frac{1}{2}} - F_{i-\frac{1}{2}}\right)}{\Delta x} \qquad (7.3)$$

where R_i is denoted the *residual*. In general,

$$F_{i+\frac{1}{2}} = F(Q^l_{i+\frac{1}{2}}, Q^r_{i+\frac{1}{2}})$$

where $Q^l_{i+\frac{1}{2}}$ and $Q^r_{i+\frac{1}{2}}$ are the left and right states reconstructed to the face at $i + \frac{1}{2}$. Thus, the residual R_i depends on

$$R_i = R_i\left(Q^l_{i+\frac{1}{2}}, Q^r_{i+\frac{1}{2}}, Q^l_{i-\frac{1}{2}}, Q^r_{i-\frac{1}{2}}\right)$$

and hence the semi-discrete form is

$$\frac{dQ_i}{dt} = R_i\left(Q^l_{i+\frac{1}{2}}, Q^r_{i+\frac{1}{2}}, Q^l_{i-\frac{1}{2}}, Q^r_{i-\frac{1}{2}}\right) \qquad (7.4)$$

7.2 Explicit Methods

There are numerous methods for explicit temporal integration. Probably the most commonly used techniques are the two- and four-stage Runge-Kutta methods (Abramowitz and Stegun, 1971; Gear, 1971; Press *et al.*, 1986). Additional methods include Bulirsch-Stoer (1970), Jameson *et al.* (1981) and Adams-Bashforth-Moulton (Press *et al.*, 1986; Gear, 1971).

7.2.1 Runge-Kutta

The *two-stage Runge-Kutta* method† is

$$
\begin{aligned}
Q_i^0 &= Q_i^n \\
Q_i^1 &= Q_i^0 + \frac{\Delta t}{2} R_i^0 \\
Q_i^2 &= Q_i^0 + \Delta t R_i^1 \\
Q_i^{n+1} &= Q_i^2
\end{aligned}
\qquad (7.5)
$$

where we introduce the temporary vectors Q_i^0, Q_i^1, and Q_i^2. The temporary vector Q_i^0 is identified as Q_i^n. The first step computes an intermediate value

† Also known as the *midpoint method*.

Q_i^1 where R_i^0 implies evaluation using Q_i^0. The second step computes the final value of Q_i^{n+1} with R_i evaluated using Q_i^1. The vector Q_i^2 is identified as Q_i^{n+1}.

The algorithm is temporally second-order, *i.e.*, the error in Q_i^{n+1} is $\mathcal{O}(\Delta t^3)$ (Exercise 7.1). Two time levels of computer storage are required (*i.e.*, Q_i^0 and an intermediate level that stores Q_i^1 and is overwritten by Q_i^2) plus separate storage for R_i. Hence, the minimum storage requirement is approximately $3^2 M$.

The *four-stage Runge-Kutta* method is

$$Q_i^0 = Q_i^n \tag{7.6}$$

$$Q_i^1 = Q_i^0 + \frac{\Delta t}{2} R_i^0 \tag{7.7}$$

$$Q_i^2 = Q_i^0 + \frac{\Delta t}{2} R_i^1 \tag{7.8}$$

$$Q_i^3 = Q_i^0 + \Delta t R_i^2 \tag{7.9}$$

$$Q_i^4 = Q_i^0 + \frac{\Delta t}{6} \left\{ R_i^0 + 2R_i^1 + 2R_i^2 + R_i^3 \right\} \tag{7.10}$$

$$Q_i^{n+1} = Q_i^4 \tag{7.11}$$

The algorithm is temporally fourth-order, *i.e.*, the error in Q_i^{n+1} is $\mathcal{O}(\Delta t^5)$ (Exercise 7.2). Two time levels of computer storage are required (*i.e.*, Q_i^0, and an intermediate level that successively stores Q_i^1, Q_i^2, Q_i^3, and Q_i^4) plus separate storage for the flux quadrature R_i at each step and for the sum of the fluxes in (7.10). Hence, the minimum storage requirement is approximately $4 \cdot 3 \cdot M$.

7.3 Implicit Methods

There are several methods for implicit temporal integration. The commonly used methods are Beam and Warming (1976) and Briley and McDonald (1973).

7.3.1 Beam-Warming

Following Beam and Warming (1976), we may expand Q_i in time as follows:

$$Q_i^{n+1} = Q_i^n + \frac{\alpha}{(1+\alpha)}\left(Q_i^n - Q_i^{n-1}\right) + \frac{\Delta t}{(1+\alpha)}\left[\left.\frac{dQ_i}{dt}\right|^n + \beta\left(\left.\frac{dQ_i}{dt}\right|^{n+1} - \left.\frac{dQ_i}{dt}\right|^n\right)\right]$$

$$(7.12)$$

where α and β are constants. It may be directly verified (Exercise 7.3) by comparison with the Taylor series expansion for Q_i that the error in this expression is

$$E = \left(\beta - \alpha - \tfrac{1}{2}\right)\mathcal{O}(\Delta t)^2 + \mathcal{O}(\Delta t)^3 \qquad (7.13)$$

Three common temporal integration methods can be defined based on the values of α and β listed in Table 7.1.

Table 7.1. *Implicit Temporal Integration Methods*

Case	Method	α	β	E
1	Trapezoidal	0	$\tfrac{1}{2}$	$\mathcal{O}(\Delta t)^3$
2	Euler implicit	0	1	$\mathcal{O}(\Delta t)^2$
3	Three point backward	$\tfrac{1}{2}$	1	$\mathcal{O}(\Delta t)^3$

Into (7.12) we substitute

$$\left.\frac{dQ_i}{dt}\right|^n = -\frac{\left(F_{i+\frac{1}{2}}^n - F_{i-\frac{1}{2}}^n\right)}{\Delta x} \qquad (7.14)$$

$$\left.\frac{dQ_i}{dt}\right|^{n+1} = -\frac{\left(F_{i+\frac{1}{2}}^{n+1} - F_{i-\frac{1}{2}}^{n+1}\right)}{\Delta x} \qquad (7.15)$$

The flux $F_{i+\frac{1}{2}}^{n+1}$ is a function of $Q_{i+\frac{1}{2}}^l$ and $Q_{i+\frac{1}{2}}^r$ evaluated at t^{n+1}. We denote

$$\left.Q_{i+\frac{1}{2}}^l\right|^{n+1} = {}^l Q_{i+\frac{1}{2}}^{n+1} \qquad (7.16)$$

and similarly for $Q_{i+\frac{1}{2}}^r$ evaluated at t^{n+1}. We therefore expand $F_{i+\frac{1}{2}}^{n+1}$ as

$$F_{i+\frac{1}{2}}^{n+1} = F_{i+\frac{1}{2}}^n + \left.\frac{\partial F}{\partial Q^l}\right|_{i+\frac{1}{2}}^n \left({}^l Q_{i+\frac{1}{2}}^{n+1} - {}^l Q_{i+\frac{1}{2}}^n\right) + \left.\frac{\partial F}{\partial Q^r}\right|_{i+\frac{1}{2}}^n \left({}^r Q_{i+\frac{1}{2}}^{n+1} - {}^r Q_{i+\frac{1}{2}}^n\right) + \mathcal{O}(\Delta t)^2$$

$$(7.17)$$

Define

$$\hat{A}_{i+\frac{1}{2}}^l = \left.\frac{\partial F}{\partial Q^l}\right|_{i+\frac{1}{2}}^n \quad \text{and} \quad \hat{A}_{i+\frac{1}{2}}^r = \left.\frac{\partial F}{\partial Q^r}\right|_{i+\frac{1}{2}}^n \qquad (7.18)$$

where it is understood that $\hat{A}^l_{i\pm\frac{1}{2}}$ and $\hat{A}^r_{i\pm\frac{1}{2}}$ are evaluated† using the data at t^n. Thus, (7.15) becomes

$$
\begin{aligned}
\frac{dQ_i}{dt}\bigg|^{n+1} = & -\frac{\left(F^n_{i+\frac{1}{2}} - F^n_{i-\frac{1}{2}}\right)}{\Delta x} + \mathcal{O}(\Delta t)^2 \\
& -\frac{1}{\Delta x}\underbrace{\left[\hat{A}^l_{i+\frac{1}{2}}\left({}^lQ^{n+1}_{i+\frac{1}{2}} - {}^lQ^n_{i+\frac{1}{2}}\right) - \hat{A}^l_{i-\frac{1}{2}}\left({}^lQ^{n+1}_{i-\frac{1}{2}} - {}^lQ^n_{i-\frac{1}{2}}\right)\right]}_{\text{terms involving left face at } i\pm\frac{1}{2}} \\
& -\frac{1}{\Delta x}\underbrace{\left[\hat{A}^r_{i+\frac{1}{2}}\left({}^rQ^{n+1}_{i+\frac{1}{2}} - {}^rQ^n_{i+\frac{1}{2}}\right) - \hat{A}^r_{i-\frac{1}{2}}\left({}^rQ^{n+1}_{i-\frac{1}{2}} - {}^rQ^n_{i-\frac{1}{2}}\right)\right]}_{\text{terms involving right face at } i\pm\frac{1}{2}}
\end{aligned}
$$

$$(7.19)$$

We now introduce the "delta" notation†

$$\delta Q^n_i = Q^{n+1}_i - Q^n_i \tag{7.20}$$

For any linear reconstruction,

$$
\begin{aligned}
\hat{A}^l_{i+\frac{1}{2}}\left({}^lQ^{n+1}_{i+\frac{1}{2}} - {}^lQ^n_{i+\frac{1}{2}}\right) - \hat{A}^l_{i-\frac{1}{2}}\left({}^lQ^{n+1}_{i-\frac{1}{2}} - {}^lQ^n_{i-\frac{1}{2}}\right) = \\
\hat{A}^l_{i+\frac{1}{2}}\delta Q^n_i - \hat{A}^l_{i-\frac{1}{2}}\delta Q^n_{i-1} + \mathcal{O}((\Delta x)^2\Delta t)
\end{aligned}
\tag{7.21}
$$

Thus, (7.12) becomes

$$
\begin{aligned}
-\nu\hat{A}^l_{i-\frac{1}{2}}\delta Q^n_{i-1} + \left[I + \nu\left(\hat{A}^l_{i+\frac{1}{2}} - \hat{A}^r_{i-\frac{1}{2}}\right)\right]\delta Q^n_i + \nu\hat{A}^r_{i+\frac{1}{2}}\delta Q^n_{i+1} = \\
\frac{\alpha}{(1+\alpha)}\delta Q^{n-1}_i - \frac{\Delta t}{(1+\alpha)}\frac{\left(F^n_{i+\frac{1}{2}} - F^n_{i-\frac{1}{2}}\right)}{\Delta x}
\end{aligned}
\tag{7.22}
$$

where

$$\nu = \frac{\beta}{(1+\alpha)}\frac{\Delta t}{\Delta x} \tag{7.23}$$

The error in this expression compared to a Taylor series expansion is

$$E = \mathcal{O}((\Delta x)^2\Delta t) + \left(\beta - \alpha - \tfrac{1}{2}\right)\mathcal{O}(\Delta t)^2 + \mathcal{O}(\Delta t)^3 \tag{7.24}$$

where the first term arises from (7.21). Equation (7.22) comprises a block-tridiagonal system of equations for integrating the Euler equations.

---

† Note that $\hat{A}^{l,r}_{i\pm\frac{1}{2}}$ are the Jacobians of the flux vector function F and thus depend on the specific choice of the flux algorithm.

† Not to be confused with (4.16).

The Jacobians $\hat{A}^{l,r}_{i\pm\frac{1}{2}}$ depend on the specific choice for the flux algorithm. For Roe's Method (Section 5.3),

$$
\begin{aligned}
\hat{A}^l_{i+\frac{1}{2}} &= \tfrac{1}{2}\left(A^l_{i+\frac{1}{2}} + |\tilde{A}|_{i+\frac{1}{2}}\right) \\
\hat{A}^r_{i+\frac{1}{2}} &= \tfrac{1}{2}\left(A^r_{i+\frac{1}{2}} - |\tilde{A}|_{i+\frac{1}{2}}\right)
\end{aligned}
$$

$$(7.25)$$

where $A^l_{i+\frac{1}{2}} = \mathcal{A}(Q^l_{i+\frac{1}{2}})$ and $\mathcal{A}^r_{i+\frac{1}{2}} = A(Q^r_{i+\frac{1}{2}})$, where $\mathcal{A}$ is the Jacobian (2.8) and the Roe matrix $|\tilde{A}|$ is treated as a constant in (7.18).

7.4 Stability of Selected Methods

In contrast to the semi-discrete stability analyses presented in Chapters 5 and 6, here we present the fully discrete stability analyses of selected explicit and implicit temporal integration methods.

7.4.1 Runge-Kutta

We consider the application of the second-order Runge-Kutta algorithm with Roe's method for the spatial fluxes

$$
\begin{aligned}
Q^0_i &= Q^n_i \\
Q^1_i &= Q^0_i + \frac{\Delta t}{2} R^0_i \\
Q^2_i &= Q^0_i + \Delta t R^1_i \\
Q^{n+1}_i &= Q^2_i
\end{aligned}
$$

with the simple first-order reconstruction

$$
\begin{aligned}
Q^l_{i+\frac{1}{2}} &= Q_i \\
Q^r_{i+\frac{1}{2}} &= Q_{i+1}
\end{aligned}
$$

We consider a von Neumann stability analysis. The residual R_i is linearized (5.46) as

$$
R_i = -\frac{1}{\Delta x}\left[A^- Q_{i+1} + (A^+ - A^-)Q_i - A^+ Q_{i-1}\right]
$$

where

$$
\begin{aligned}
A^+ &= \tfrac{1}{2}(A + |A|) &= T\Lambda^+ T^{-1} \\
A^- &= \tfrac{1}{2}(A - |A|) &= T\Lambda^- T^{-1}
\end{aligned}
$$

$$(7.26)$$

and

$$\begin{aligned}
\Lambda^+ &= \text{diag}\{\max(\lambda_m, 0)\} \\
\Lambda^- &= \text{diag}\{\min(\lambda_m, 0)\}
\end{aligned}$$

where λ_m for $m = 1, 2, 3$ are the eigenvalues (2.12).

Substituting into the first Runge-Kutta step,

$$Q_i^1 = Q_i^n - \frac{\Delta t}{2\Delta x}\left[A^- Q_{i+1}^n + (A^+ - A^-)Q_i^n - A^+ Q_{i-1}^n\right] \tag{7.27}$$

Similarly, substituting into the second Runge-Kutta step,

$$Q_i^{n+1} = Q_i^n - \frac{\Delta t}{\Delta x}\left[A^- Q_{i+1}^1 + (A^+ - A^-)Q_i^1 - A^+ Q_{i-1}^1\right] \tag{7.28}$$

Substituting (7.27) into (7.28) yields†

$$\begin{aligned}
Q_i^{n+1} &= \frac{\alpha^2}{2}A^- A^- Q_{i+2}^n + \left[\alpha^2(A^+ - A^-)A^- - \alpha A^-\right]Q_{i+1}^n \\
&\quad + \left[I - \alpha(A^+ - A^-) + \frac{\alpha^2}{2}(A^+ A^+ + A^- A^- - 4A^+ A^-)\right]Q_i^n \\
&\quad + \left[-\alpha^2(A^+ - A^-)A^+ + \alpha A^+\right]Q_{i-1}^n \\
&\quad + \frac{\alpha^2}{2}A^+ A^+ Q_{i-2}^n
\end{aligned} \tag{7.29}$$

where‡

$$\alpha = \frac{\Delta t}{\Delta x} \tag{7.30}$$

The solution Q_i is expanded as a Fourier series as in Section 5.3.2. The Fourier components satisfy

$$\hat{Q}_k^{n+1} = G\hat{Q}_k^n$$

where the amplification matrix G is

$$\begin{aligned}
G &= \frac{\alpha^2}{2}e^{\iota 2k\Delta x}A^- A^- + e^{\iota k\Delta x}\left[\alpha^2(A^+ - A^-)A^- - \alpha A^-\right] \\
&\quad + \left[I - \alpha(A^+ - A^-) + \frac{\alpha^2}{2}(A^+ A^+ + A^- A^- - 4A^+ A^-)\right] \\
&\quad + e^{-\iota k\Delta x}\left[-\alpha^2(A^+ - A^-)A^+ + \alpha A^+\right]
\end{aligned}$$

† Note that $A^+ A^- = A^- A^+$, $A^+|A| = |A|A^+$, and $A^-|A| = |A|A^-$ based on the structure of these matrices.
‡ Note that α in (7.30) is not the same as α in (7.12).

$$+ e^{-\iota 2k\Delta x}\frac{\alpha^2}{2}A^+ A^+ \tag{7.31}$$

It is evident that G can be written in the form

$$G = TDT^{-1}$$

where D is the diagonal matrix

$$
\begin{aligned}
D \;=\;& \frac{\alpha^2}{2}e^{\iota 2k\Delta x}\Lambda^-\Lambda^- + e^{\iota k\Delta x}\left[\alpha^2(\Lambda^+ - \Lambda^-)\Lambda^- - \alpha\Lambda^-\right] \\
& + \left[I - \alpha(\Lambda^+ - \Lambda^-) + \frac{\alpha^2}{2}(\Lambda^+\Lambda^+ + \Lambda^-\Lambda^- - 4\Lambda^+\Lambda^-)\right] \\
& + e^{-\iota k\Delta x}\left[-\alpha^2(\Lambda^+ - \Lambda^-)\Lambda^+ + \alpha\Lambda^+\right] \\
& + e^{-\iota 2k\Delta x}\frac{\alpha^2}{2}\Lambda^+\Lambda^+
\end{aligned}
\tag{7.32}
$$

Defining

$$
\begin{aligned}
\lambda_m^+ &= \max(\lambda_m, 0) \\
\lambda_m^- &= \min(\lambda_m, 0)
\end{aligned}
\tag{7.33}
$$

it is evident that

$$
\begin{aligned}
\lambda_{G_m} \;=\;& \frac{\alpha^2}{2}e^{\iota 2k\Delta x}\lambda_m^-\lambda_m^- + e^{\iota k\Delta x}\left[\alpha^2(\lambda_m^+ - \lambda_m^-)\lambda_m^- - \alpha\lambda_m^-\right] \\
& + \left[I - \alpha(\lambda_m^+ - \lambda_m^-) + \frac{\alpha^2}{2}(\lambda_m^+\lambda_m^+ + \lambda_m^-\lambda_m^- - 4\lambda_m^+\lambda_m^-)\right] \\
& + e^{-\iota k\Delta x}\left[-\alpha^2(\lambda_m^+ - \lambda_m^-)\lambda_m^+ + \alpha\lambda_m^+\right] \\
& + e^{-\iota 2k\Delta x}\frac{\alpha^2}{2}\lambda_m^+\lambda_m^+
\end{aligned}
\tag{7.34}
$$

This can be compared with the result for the semi-discrete stability analysis in (5.53):

$$\lambda_{G_m} = -\frac{1}{\Delta x}\left[e^{\iota k\Delta x}\lambda_m^- + \lambda_m^+ - \lambda_m^- - e^{-\iota k\Delta x}\lambda_m^+\right]$$

Define

$$a = \alpha\lambda^+ = \frac{\Delta t\lambda^+}{\Delta x}$$

$$b = \alpha\lambda^- = \frac{\Delta t\lambda^-}{\Delta x} \tag{7.35}$$

where the subscript m is omitted but implied. Writing

$$\lambda_{G_m} = \omega_{r_m} + \iota\omega_{\iota_m}$$

we have

$$\omega_{r_m} = \frac{(a^2 + b^2)}{2} \cos 2\theta + (a - b)[1 - (a - b)] \cos \theta$$
$$+ 1 - (a - b) + \tfrac{1}{2}(a - b)^2 - ab$$
$$\omega_{\iota_m} = \frac{(b^2 - a^2)}{2} \sin 2\theta - (a + b)[1 - (a - b)] \sin \theta$$

where

$$\theta = k\Delta x$$

There are three possible cases, which are shown in Table 7.2.

Table 7.2. *Cases*

Case	a	b
$\lambda_m > 0$	> 0	0
$\lambda_m < 0$	0	< 0
$\lambda_m = 0$	0	0

For $\lambda_m > 0$,

$$\omega_{r_m} = \frac{a^2}{2} \cos 2\theta + a(1 - a) \cos \theta + 1 - a + \frac{a^2}{2}$$
$$\omega_{\iota_m} = -\frac{a^2}{2} \sin 2\theta - a(1 - a) \sin \theta \qquad (7.36)$$

For $\lambda_m < 0$,

$$\omega_{r_m} = \frac{b^2}{2} \cos 2\theta - b(1 + b) \cos \theta + 1 + b + \frac{b^2}{2}$$
$$\omega_{\iota_m} = \frac{b^2}{2} \sin 2\theta - b(1 + b) \sin \theta \qquad (7.37)$$

For $\lambda_m = 0$,

$$\omega_{r_m} = 1$$
$$\omega_{\iota_m} = 0 \qquad (7.38)$$

Letting $\theta = -\tilde{\theta}$ and $b = -\tilde{b}$, it is evident that the second case is equivalent to the first.

Stability (3.81) requires

$$|\lambda_{G_m}| \leq 1 \qquad (7.39)$$

Clearly, the case $\lambda_m = 0$ satisfies this condition. Thus, it is necessary to determine the conditions for which $\lambda_m > 0$ satisfy this condition. Equations (7.36) represent a parametric curve in the complex λ_G plane with the parameter θ. This curve can be rewritten as

$$\left\{ \left[\frac{\omega_r - (1-a)}{a} \right]^2 + \left[\frac{\omega_\iota}{a} \right]^2 - [\omega_r - (1-a)] \right\}^2 =$$
$$(1-a)^2 \left\{ \left[\frac{\omega_r - (1-a)}{a} \right]^2 + \left[\frac{\omega_\iota}{a} \right]^2 \right\} \qquad (7.40)$$

This is the *Limaçon of Pascal*. It is shown in Fig. 7.1 for $a = 0.8$.

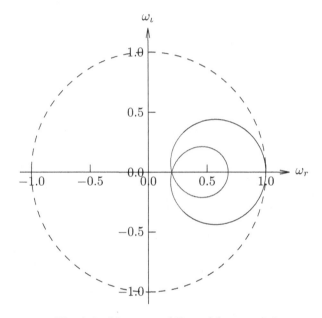

Fig. 7.1. Limaçon of Pascal for $a = 0.8$

The condition for stability in (7.39) yields

$$a \leq 1 \qquad (7.41)$$

and thus

$$\frac{\Delta t |\lambda_m^\pm|}{\Delta x} \leq 1$$

where $\lambda_m^\pm$ indicates consideration of both λ_m^+ and λ_m^-. This may be rewritten as

$$\Delta t \leq \Delta t_{CFL}$$

where

$$\Delta t_{CFL} = \min_i \frac{\Delta x}{|\lambda_m^{\pm}|}$$

or

$$\mathcal{C} \leq 1$$

where $\mathcal{C}$ is the Courant number:

$$\mathcal{C} = \frac{\Delta t}{\Delta t_{CFL}}$$

7.4.2 Beam-Warming

We consider the application of the Beam-Warming algorithm with Roe's method for the spatial fluxes. We assume $\alpha = 0$ for simplicity (*i.e.*, trapezoidal rule or Euler implicit time integration). We treat A and $|\tilde{A}|$ as constants. The Roe matrix $|\tilde{A}|$ may be expanded as

$$|\tilde{A}| = |A| + \mathcal{O}(\Delta Q_i) \tag{7.42}$$

Thus,

$$A_{i+\frac{1}{2}}^l = \tfrac{1}{2}(A + |A|) = T\Lambda^+ T$$
$$A_{i+\frac{1}{2}}^r = \tfrac{1}{2}(A - |A|) = T\Lambda^- T$$

where

$$\Lambda^+ = \text{diag}\,\{\max(\lambda_m, 0)\}$$
$$\Lambda^- = \text{diag}\,\{\min(\lambda_m, 0)\}$$

We denote

$$A^+ = T\Lambda^+ T$$
$$A^- = T\Lambda^- T$$

Beam-Warming's algorithm (7.22) becomes

$$-\nu A^+ \delta Q_{i-1}^n + \left[I + \nu\left(A^+ - A^-\right)\right]\delta Q_i^n + \nu A^- \delta Q_{i+1}^n =$$
$$-\frac{\Delta t}{\Delta x}\left[A^- Q_{i+1}^n + \left(A^+ - A^-\right)Q_i^n - A^+ Q_{i-1}^n\right] \tag{7.43}$$

The solution Q_i is expanded in a Fourier series as in Section 5.3.2:

$$Q(x, t) = \sum_{l=-N+1}^{l=N} \hat{Q}_k(t) e^{\iota k x}$$

where $k = 2\pi/l$. We denote $\hat{Q}_k^n = \hat{Q}_k(t^n)$. Substitution into (7.43) yields

$$G_l \hat{Q}_k^{n+1} = G_r \hat{Q}_k^n \qquad (7.44)$$

where

$$
\begin{aligned}
G_l &= -\nu e^{-\iota k \Delta x} A^+ + \left[I + \nu\left(A^+ - A^-\right)\right] + \nu e^{+\iota k \Delta x} A^- \\
G_r &= \mu e^{-\iota k \Delta x} A^+ + \left[I - \mu\left(A^+ - A^-\right)\right] - \mu e^{+\iota k \Delta x} A^-
\end{aligned}
\qquad (7.45)
$$

and

$$
\begin{aligned}
\nu &= \beta \frac{\Delta t}{\Delta x} \\
\mu &= (1 - \beta)\frac{\Delta t}{\Delta x}
\end{aligned}
\qquad (7.46)
$$

The matrices D_l and D_r can be factored as

$$
\begin{aligned}
G_l &= T D_l T^{-1} \\
G_r &= T D_r T^{-1}
\end{aligned}
\qquad (7.47)
$$

where

$$
\begin{aligned}
D_l &= -\nu e^{-\iota k \Delta x}\Lambda^+ + \left[I + \nu\left(\Lambda^+ - \Lambda^-\right)\right] + \nu e^{+\iota k \Delta x}\Lambda^- \\
D_r &= \mu e^{-\iota k \Delta x}\Lambda^+ + \left[I - \mu\left(\Lambda^+ - \Lambda^-\right)\right] - \mu e^{+\iota k \Delta x}\Lambda^-
\end{aligned}
\qquad (7.48)
$$

The matrices D_l and D_r are diagonal and may be written as

$$
\begin{aligned}
D_l &= \operatorname{diag}\left\{\lambda_{D_{l_m}}\right\} \\
D_r &= \operatorname{diag}\left\{\lambda_{D_{r_m}}\right\}
\end{aligned}
\qquad (7.49)
$$

where

$$
\begin{aligned}
\lambda_{D_{l_m}} &= -\nu e^{-\iota k \Delta x}\lambda_m^+ + \left[1 + \nu\left(\lambda_m^+ - \lambda_m^-\right)\right] + \nu e^{+\iota k \Delta x}\lambda_m^- \\
\lambda_{D_{r_m}} &= \mu e^{-\iota k \Delta x}\lambda_m^+ + \left[1 - \mu\left(\lambda_m^+ - \lambda_m^-\right)\right] - \mu e^{+\iota k \Delta x}\lambda_m^-
\end{aligned}
\qquad (7.50)
$$

Equation (7.44) becomes

$$D_l \tilde{Q}_k^{n+1} = D_r \tilde{Q}_k^n \qquad (7.51)$$

where

$$\tilde{Q}_k^n = T^{-1}\hat{Q}_k^n \qquad (7.52)$$

Defining

$$
\tilde{Q}_k = \left\{
\begin{array}{c}
\tilde{Q}_{k_1} \\
\tilde{Q}_{k_2} \\
\tilde{Q}_{k_3}
\end{array}
\right\}
\qquad (7.53)
$$

the solution is then

$$\tilde{Q}_{km}^{n+1} = \frac{\lambda_{D_{rm}}}{\lambda_{D_{lm}}} \tilde{Q}_{km}^n \tag{7.54}$$

The necessary condition for stability is therefore

$$|\lambda_{D_{rm}}| \le |\lambda_{D_{lm}}| \quad \text{for} \quad m = 1, 2, 3 \tag{7.55}$$

Now

$$
\begin{aligned}
\lambda_{D_{lm}} &= 1 + \nu \left(\lambda_m^+ - \lambda_m^-\right)(1 - \cos\theta) + \iota\nu \left(\lambda_m^+ + \lambda_m^-\right)\sin\theta \\
\lambda_{D_{rm}} &= 1 - \mu \left(\lambda_m^+ - \lambda_m^-\right)(1 - \cos\theta) - \iota\mu \left(\lambda_m^+ + \lambda_m^-\right)\sin\theta \quad (7.56)
\end{aligned}
$$

where $\theta = k\Delta x$. The stability condition (7.55) then becomes

$$
\begin{aligned}
\left(\lambda_m^+ - \lambda_m^-\right)^2 (1 - \cos\theta)^2 \left(\frac{\Delta t}{\Delta x}\right)^2 (2\beta - 1) & \\
+ \left(\lambda_m^+ - \lambda_m^-\right)(1 - \cos\theta)\, 2\frac{\Delta t}{\Delta x} & \\
+ \left(\lambda_m^+ + \lambda_m^-\right)^2 \sin^2\theta \left(\frac{\Delta t}{\Delta x}\right)^2 (2\beta - 1) \ \geq \ 0 & \tag{7.57}
\end{aligned}
$$

Now $\lambda_m^+ - \lambda_m^- \ge 0$ and therefore the second term is nonnegative for all θ (*i.e.*, for all $k\Delta x$). The first and third terms are nonnegative for $\beta \ge \frac{1}{2}$. Thus, the necessary condition for stability is satisfied without any constraint on Δt.

Exercises

7.1   Prove that the two-stage Runge-Kutta method (7.5) is second-order accurate.

SOLUTION
From (7.5),

$$Q_i^2 = Q_i^0 + \Delta t R_i^1$$

where R_i^1, defined in (7.3), is a function of the Q_j^1 in a neighborhood about i based on the reconstruction method. For example, if the reconstruction (4.15) is used, then R_i^1 depends on $Q_{i-2}^0, Q_{i-1}^0, Q_i^0, Q_{i+1}^0$, and Q_{i+2}^0. We write

$$R_i^1 = R_i(Q_j^1)$$

and thus

$$R_i^1 = R_i(Q_j^0 + \tfrac{\Delta t}{2} R_j^0)$$

Expanding in a Taylor series,

$$R_i^1 = R_i(Q_j^0) + \sum_j \frac{\partial R_i}{\partial Q_j} \frac{\Delta t}{2} R_j^0 + \mathcal{O}(\Delta t)^2$$

Thus,

$$Q_i^2 = Q_i^0 + \Delta t R_i(Q_j^0) + \sum_j \frac{\partial R_i}{\partial Q_j} \frac{(\Delta t)^2}{2} R_j^0 + \mathcal{O}(\Delta t)^3$$

Now

$$\frac{d^2 Q_i}{dt^2} = \frac{dR_i}{dt} = \sum_j \frac{\partial R_i}{\partial Q_j} \frac{dQ_j}{dt} = \sum_j \frac{\partial R_i}{\partial Q_j} R_j$$

and using

$$\frac{dQ_i}{dt} = R_i$$

then

$$Q_i^2 = Q_i^0 + \frac{dQ_i}{dt} \Delta t + \frac{d^2 Q_i}{dt^2} \frac{(\Delta t)^2}{2} + \mathcal{O}(\Delta t)^3$$

Thus, the two-stage Runge-Kutta method is second-order accurate.

7.2 Prove that the four-stage Runge-Kutta method (7.11) is fourth-order accurate

7.3 Derive the expression (7.13) for the truncation error of the expansion (7.12).

SOLUTION

Consider Case 1 in Table 7.1. The expansion (7.12) is

$$Q_i^{n+1} = Q_i^n + \frac{\Delta t}{2} \left(\left.\frac{dQ_i}{dt}\right|^n + \left.\frac{dQ_i}{dt}\right|^{n+1} \right)$$

Expanding

$$\left.\frac{dQ_i}{dt}\right|^{n+1} = \left.\frac{dQ_i}{dt}\right|^n + \left.\frac{d^2 Q_i}{dt^2}\right|^n \Delta t + \left.\frac{d^3 Q_i}{dt^3}\right|^n \frac{(\Delta t)^2}{2} + \mathcal{O}(\Delta t)^3$$

Substituting into (7.12),

$$Q_i^{n+1} = Q_i^n + \left.\frac{dQ_i}{dt}\right|^n \Delta t + \left.\frac{d^2 Q_i}{dt^2}\right|^n \frac{(\Delta t)^2}{2} + \left.\frac{d^3 Q_i}{dt^3}\right|^n \frac{(\Delta t)^3}{4} + \mathcal{O}(\Delta t)^4$$

This expression agrees with the Taylor series expansion up to and including $\mathcal{O}(\Delta t)^2$. Thus the error $E = \mathcal{O}(\Delta t)^3$. The error for Cases 2 and 3 may be found in a similar manner.

7.4 Prove that the error in (7.17) is $\mathcal{O}(\Delta t)^2$.

7.5 Prove that the error in (7.19) is $\mathcal{O}(\Delta t)^2$.

SOLUTION

From (7.17), the error in the expansion for $F_{i+\frac{1}{2}}^{n+1}$ is $\mathcal{O}(\Delta t)^2$, specifically,

$$F_{i+\frac{1}{2}}^{n+1} = F_{i+\frac{1}{2}}^n + \hat{A}_{i+\frac{1}{2}}^l \left({}^l Q_{i+\frac{1}{2}}^{n+1} - {}^l Q_{i+\frac{1}{2}}^n \right) + \hat{A}_{i+\frac{1}{2}}^r \left({}^r Q_{i+\frac{1}{2}}^{n+1} - {}^r Q_{i+\frac{1}{2}}^n \right) + \mathcal{O}(\Delta t)^2$$

Thus, the error in expansion for $F_{i+\frac{1}{2}}^{n+1} - F_{i-\frac{1}{2}}^{n+1}$ is $\mathcal{O}(\Delta x (\Delta t)^2)$. Substituting into (7.15), the error in (7.19) is $\mathcal{O}(\Delta t)^2$.

7.6 Prove that the error in (7.21) is $\mathcal{O}((\Delta x)^2 \Delta t)$.

7.7   Derive the expression (7.25) for the matrices $\hat{A}^l_{i+\frac{1}{2}}$ and $\hat{A}^r_{i+\frac{1}{2}}$.

SOLUTION

The flux expression for Roe's Method in (5.39) is

$$F_{i+\frac{1}{2}} = \tfrac{1}{2}\left[F_l + F_r + \tilde{S}|\tilde{\Lambda}|\tilde{S}^{-1}(Q_l - Q_r)\right]$$

Thus,

$$\frac{\partial F}{\partial Q^l} = \tfrac{1}{2}\left[\frac{\partial F_l}{\partial Q^l} + \tilde{S}|\tilde{\Lambda}|\tilde{S}^{-1}\right]$$

where the Roe matrix $\tilde{S}|\tilde{\Lambda}|\tilde{S}^{-1}$ is treated as constant. Thus,

$$\frac{\partial F}{\partial Q^l} = \tfrac{1}{2}\left[A^l + |\tilde{A}|\right]$$

Similarly,

$$\frac{\partial F}{\partial Q^r} = \tfrac{1}{2}\left[A^r - |\tilde{A}|\right]$$

7.8   Prove that the Limaçon of Pascal lies within or on the unit circle if (7.41) is satisfied.

7.9   Prove that the error in the expansion (7.42) is $\mathcal{O}(\Delta Q_i)$.

SOLUTION

The expansion is

$$|\tilde{A}| = |A| + \mathcal{O}(\Delta Q_i)$$

Now

$$|\tilde{A}| = \tilde{S}|\tilde{\Lambda}|\tilde{S}^{-1}$$

We rewrite the above equation as

$$|\tilde{A}| = \left[T + (\tilde{S} - T)\right]\left[|\Lambda| + (|\tilde{\Lambda}| - |\Lambda|)\right]\left[T^{-1} + (\tilde{S}^{-1} - T^{-1})\right]$$

Thus,

$$\begin{aligned}
|\tilde{A}| = \quad & T|\Lambda|T^{-1} \\
& + T|\Lambda|(\tilde{S}^{-1} - T^{-1}) + T(|\tilde{\Lambda}| - |\Lambda|)T^{-1} + (\tilde{S} - T)|\Lambda|T^{-1} \\
& + \text{terms that are quadratic in the differences} \qquad\qquad \text{(E7.1)}
\end{aligned}$$

Now, from (2.17) and (5.28), we have

$$\tilde{S} - T = \mathcal{O}(\Delta Q_i)$$

and similarly for the other differences. Thus,

$$|\tilde{A}| = |A| + \mathcal{O}(\Delta Q_i)$$

7.10   Derive the stability result for Beam-Warming with $\alpha = \frac{1}{2}$.

8

TVD Methods

The theoretical basis of TVD methods is sound for scalar one-dimensional problems only. In practical, non-linear, multi-dimensional problems, the accumulated experience of numerous applications has demonstrated that the one-dimensional scalar theory serves well as a guideline for extending the ideas, on a more or less empirical basis.

Eleuterio F. Toro (1997)

8.1 Introduction

It is instructive to examine whether a proper reconstruction method (Chapter 4) with a Godunov (Chapter 5) or flux vector split algorithm (Chapter 6) is capable of computing oscillation-free solutions of the Euler equations in the presence of discontinuities (*i.e.*, shock waves or contact surfaces). We consider the problem of Section 2.8 with $\epsilon = 0.1$ for a domain $0 < \kappa x < 2\pi$ for which a shock wave forms at $\kappa a_o t_s = 8.333$. We first consider the Godunov algorithm with a third-order ($\kappa = \frac{1}{3}$) reconstruction (4.19) with no limiter. Figure 8.1 displays the velocity profile at $\kappa a_o t = 14$. Oscillations have formed in the vicinity of the shock. These oscillations are unphysical (and therefore undesirable) and are attributable to the numerical algorithm.

It seems likely that the oscillations in the vicinity of the shock in Fig. 8.1 are attributable to the reconstruction algorithm, since no limiter was employed in Fig. 8.1 to prevent oscillations in the reconstructed variables (see Fig. 4.6). Therefore, we consider the same algorithm but with the minmod limiter in reconstruction. Figure 8.2 displays the velocity profile at

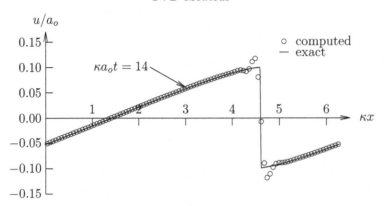

Fig. 8.1. Velocity at $\kappa a_o t = 14$ using Godunov's Method, a third-order reconstruction, and no limiter

$\kappa a_o t = 14$. The magnitudes of the oscillations have been reduced significantly, but on closer inspection it is evident that the oscillations have not been entirely eliminated.

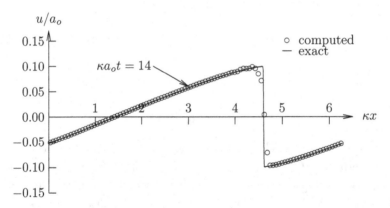

Fig. 8.2. Velocity at $\kappa a_o t = 14$ for Godunov with a third-order reconstruction and min-mod limiter

It is therefore evident that the elimination of unphysical oscillations in the reconstruction does *not* guarantee the elimination of unphysical oscillations in the computed solution of the Euler equations. However, the elimination of unphysical oscillations in the reconstruction through limiters would be expected to *reduce* the magnitude of the unphysical oscillations in the computed solution of the Euler equations. Nevertheless, the elimination of *all* unphysical oscillations in the computed solution of the Euler equations requires modification of the fluxes, since the evolution of the solution $Q_i(t)$ is directly a consequence of the integral of the fluxes across the control volume according to (3.9):

$$\frac{dQ_i}{dt} + \frac{\left(F_{i+\frac{1}{2}} - F_{i-\frac{1}{2}}\right)}{\Delta x} = 0$$

8.2 Total Variation

Harten (1983,1984) introduced the concept of Total Variation (TV) of a function as both a means of proving the convergence of a discrete system of equations to their exact solution and as a technique for quantifying the extent of oscillations in the discrete solution. Consider the computed solution in Fig. 8.3. The Total Variation of the velocity u_i is defined as

$$TV(u) = \sum_i |u_i - u_{i-1}| \qquad (8.1)$$

The Total Variation of other flow variables can be similarly defined. For the periodic smooth profile shown in Fig. 8.3,

$$TV(u) = 2(\max u_i - \min u_i) \qquad (8.2)$$

The appearance of oscillations in a solution typically increases the TV. It is straightforward to show (Exercise 8.1) for a periodic function u that

$$TV(u) = 2\sum (u_{\max_i} - u_{\min_i}) \qquad (8.3)$$

where $u_{\max_i}$ and $u_{\min_i}$ are the local maxima and minima of u. Thus, the appearance of oscillations in the solution increases the number of terms in the above expression and typically increases the value of TV.

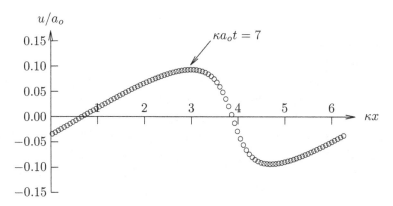

Fig. 8.3. Computed solution using Van Leer's Method with a first-order reconstruction

The TV concept can be extended to a general function $u(x)$ according to[†]

$$TV(u) = \lim_{\delta \to 0} \sup \frac{1}{\delta} \int_{-\infty}^{\infty} |u(x + \delta) - u(x)| \, dx \tag{8.4}$$

provided that $TV(u)$ is finite. For a smooth function, this becomes

$$TV(u) = \int_{-\infty}^{\infty} \left| \frac{\partial u}{\partial x} \right| \, dx \tag{8.5}$$

provided $\partial u / \partial x$ approaches zero sufficiently fast as $|x|$ approaches infinity.

Harten (1984), Boris and Book (1976), and others introduced algorithms for a hyperbolic system of equations that have a nonincreasing value of TV in time. These methods are denoted *Total Variation Diminishing (TVD)*.[†] A numerical algorithm for a scalar function u is TVD if

$$TV(u^{n+1}) \leq TV(u^n) \tag{8.6}$$

A related concept is *Total Variation Bounded (TVB)*. A numerical algorithm is TVB if

$$TV(u^n) \leq B \quad \text{for} \quad 0 \leq t \leq \tau \tag{8.7}$$

for all n such that $n\delta t \leq \tau$, where B depends only on $TV(u^o)$ and δt satisfies a Courant-Friedrichs-Lewy condition if applicable.

We examine the Total Variation of the solution of the problem described in Section 2.8. The initial condition is defined by (2.92) with $\epsilon = 0.1$, and the domain $0 < \kappa x < 2\pi$ is discretized into 100 uniform cells. Periodic boundary conditions are imposed on the left and right boundaries. The timestep is defined by (5.8) with $\mathcal{C} = 0.46$, and the second-order Runge-Kutta algorithm (7.5) is employed.

First, we consider the first-order reconstruction (5.1). Figure 8.4 displays the Total Variation of ρ, ρu, and ρe normalized by ρ_o, $\rho_o a_o$, and $\rho_o a_o^2$, respectively, where ρ_o and a_o are the undisturbed density and speed of sound, respectively. The flux algorithm is Roe's method. The Total Variation is seen to decrease for all three variables. Thus, the algorithm is TVD; however, the Total Variation is significantly below the exact value, implying excessive numerical dissipation. The velocity profile at $t = 14$ is shown in Fig. 8.5. Considerable diffusion and decay of the profile in the vicinity of the shock is evident, which is attributable to the use of first-order reconstruction.

---

† The supremum (denoted sup) of a set is the least upper bound of the set (Ahlfors, 1953).
† A more correct nomenclature would be *Total Variation Non Increasing (TVNI)*.

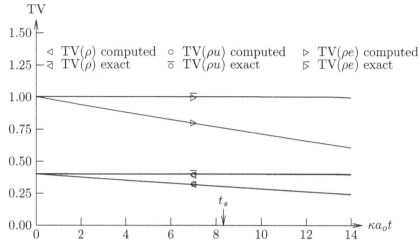

Fig. 8.4. Total Variation of ρ, ρu, and ρe

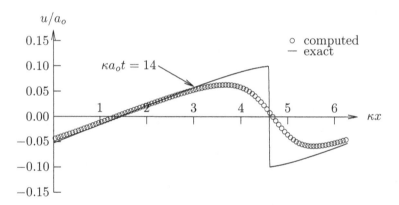

Fig. 8.5. Velocity at $\kappa a_o t = 14$

The Total Variation of ρu is shown in Fig. 8.6 for five different flux algorithms using first-order reconstruction. Notwithstanding the quantitative differences, it is evident that all methods are TVD and underpredict the exact solution.

Second, we consider the third-order upwind-biased reconstruction (4.19) with $\kappa = \frac{1}{3}$ but no reconstruction limiter. The flux algorithm is Godunov's method (*i.e.*, the exact solution of the General Riemann Problem is employed for the flux.) The behavior of $TV(\rho u)$, shown in Fig. 8.7, is essentially constant up to the time of shock formation but then increases in contrast to the exact solution. The resulting velocity profile at $\kappa a_o t = 14$, shown in

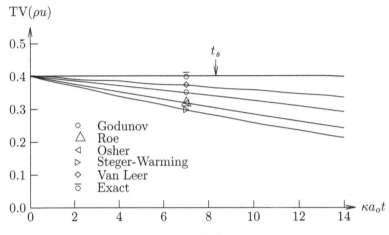

Fig. 8.6. TV(ρu) for different flux algorithms

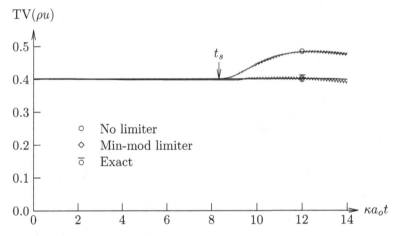

Fig. 8.7. TV(ρu) for Godunov with a third-order reconstruction with and without reconstruction limiter

Fig. 8.1, displays oscillations on both sides of the shock, which accounts for the increase in TV(ρu).

Third, we consider the third-order upwind-biased reconstruction with the min-mod limiter. The flux algorithm is Godunov's Method. The TV(ρu), shown in Fig. 8.7, is significantly better than the previous solution, which did not employ a limiter. The velocity profile in Fig. 8.2 displays only a slight oscillation in the vicinity of the shock. We conclude that the reconstruction limiter has virtually eliminated the oscillations but that the method is not precisely TVD.

Toro (1997) classifies TVD methods into two categories. The first category is *flux limiters*. This approach is based on a dynamic combination of high- and low-order expressions for the flux. The second category is *slope limiters*. This approach imposes a limit on the reconstruction to achieve TVD behavior. In the following sections, we present one example of each of these two approaches. Additional examples are presented in Toro (1997).

8.3 Flux Corrected Transport

Boris and Book (1973) introduced a flux limiter method that combines a low-order and high-order flux algorithm to achieve a TVD scheme. The algorithm is further described in Book *et al.* (1975), Boris and Book (1976), and Zalesak (1979).

8.3.1 Algorithm

Consider the semi-discrete form (3.9) of the Euler equations

$$\frac{dQ_i}{dt} + \frac{\left(F_{i+\frac{1}{2}} - F_{i-\frac{1}{2}}\right)}{\Delta x} = 0 \tag{8.8}$$

For simplicity, we utilize first-order Euler integration of (8.8) in time and denote the solution at time t^n by Q_i^n.

The algorithm is composed of two steps. In the first step, an intermediate solution† $\hat{Q}_i$ is generated using (typically) a first-order (*i.e.*, *low* order‡) flux algorithm $F_{i+\frac{1}{2}}^l$:

$$\hat{Q}_i = Q_i^n - \frac{\Delta t}{\Delta x}\left(F_{i+\frac{1}{2}}^l - F_{i-\frac{1}{2}}^l\right) \tag{8.9}$$

The flux algorithm $F_{i\pm\frac{1}{2}}^l$ is chosen to satisfy the maximum principle, *i.e.*, the solution $\hat{Q}_i$ does not exceed the range of the initial data. This intermediate solution $\hat{Q}_i$ exhibits significant diffusion that leads to an excessive broadening of discontinuities.

In the second step, a corrected flux $F_{i+\frac{1}{2}}^c$, is employed, which is given by

$$F_{i+\frac{1}{2}}^c = \alpha_{i+\frac{1}{2}} F_{i+\frac{1}{2}}^a \tag{8.10}$$

---

† Referred to as the *transported and diffused* solution (Zalesak, 1979).
‡ Hence the superscript *l*.

where $F^a_{i+\frac{1}{2}}$ is an *anti-diffusive flux*[1] given by

$$F^a_{i+\frac{1}{2}} = F^h_{i+\frac{1}{2}} - F^l_{i+\frac{1}{2}} \tag{8.11}$$

where $F^h_{i+\frac{1}{2}}$ is a higher order flux (*i.e.*, generated using a higher order reconstruction) and $\alpha_{i+\frac{1}{2}}$ is a coefficient (denoted the *hybridization parameter*) determined by the requirement that the updated solution given by

$$Q^{n+1}_i = \hat{Q}_i - \frac{\Delta t}{\Delta x}\left(F^c_{i+\frac{1}{2}} - F^c_{i-\frac{1}{2}}\right) \tag{8.12}$$

does not introduce any new extrema relative to the set $(\hat{Q}_{i-2}, \ldots, \hat{Q}_{i+2})$. The anti-diffusive flux is formed using the intermediate solution.

Boris and Book specify the following expression for the corrected flux:

$$F^c_{i+\frac{1}{2}} = S_{i+\frac{1}{2}} \max\left\{0, K_{i+\frac{1}{2}}\right\} \tag{8.13}$$

where

$$K_{i+\frac{1}{2}} = \min\left\{\left|F^a_{i+\frac{1}{2}}\right|, S_{i+\frac{1}{2}}\Delta\check{Q}_{i+2}, S_{i+\frac{1}{2}}\Delta\check{Q}_i\right\} \tag{8.14}$$

with

$$\Delta\hat{Q}_i = \hat{Q}_i - \hat{Q}_{i-1} \tag{8.15}$$

and

$$\Delta\check{Q}_i = \frac{\Delta x}{\Delta t}\Delta\hat{Q}_i \tag{8.16}$$

and

$$S_{i+\frac{1}{2}} = \mathrm{sign}\left(F^a_{i+\frac{1}{2}}\right) \tag{8.17}$$

The algorithm in (8.9) to (8.17) is *Flux Corrected Transport (FCT)*. From (8.12) to (8.14), it is evident that Q^{n+1}_i depends on the six quantities

$$\Delta\hat{Q}_{i-1}, \quad \Delta\hat{Q}_i, \quad \Delta\hat{Q}_{i+1}, \quad \Delta\hat{Q}_{i+2}, \quad F^a_{i-\frac{1}{2}}, \quad F^a_{i+\frac{1}{2}},$$

There are 64 (*i.e.*, 2^6) possible cases to evaluate for determining the effect of FCT on Q^{n+1}_i, corresponding to the two possible values of the signs of $\Delta\hat{Q}_{i-1}, \Delta\hat{Q}_i, \Delta\hat{Q}_{i+1}, \Delta\hat{Q}_{i+2}, F^a_{i-\frac{1}{2}}$, and $F^a_{i-\frac{1}{2}}$. These cases may be divided into 16 classes based on the signs of $\Delta\hat{Q}_{i-1}, \Delta\hat{Q}_i, \Delta\hat{Q}_{i+1}, \Delta\hat{Q}_{i+1}$ as indicated in Table 8.1.

We establish three important properties of FCT. First, it may be shown (Exercise 8.7) that

$$\mathrm{sign}\left(F^c_{i+\frac{1}{2}}\right) = \mathrm{sign}\left(F^a_{i+\frac{1}{2}}\right) \quad \text{if} \quad F^c_{i+\frac{1}{2}} \neq 0 \tag{8.18}$$

Table 8.1. *Classes for Flux Corrected Transport*

Class	$\Delta\hat{Q}_{i-1}$	$\Delta\hat{Q}_i$	$\Delta\hat{Q}_{i+1}$	$\Delta\hat{Q}_{i+2}$
1	+	+	+	+
2	+	+	+	−
3	+	+	−	+
4	+	−	+	+
5	−	+	+	+
6	+	+	−	−
7	+	−	+	−
8	−	+	+	−
9	+	−	−	+
10	−	+	−	+
11	−	−	+	+
12	+	−	−	−
13	−	+	−	−
14	−	−	+	−
15	−	−	−	+
16	−	−	−	−

Thus, the corrected flux $F^c_{i+\frac{1}{2}}$ (if it is nonzero) has the same character as the anti-diffusive flux $F^a_{i+\frac{1}{2}}$. Second, it may be shown (Exercise 8.8) that

$$F^c_{i+\frac{1}{2}} \neq 0 \quad \text{iff} \quad \text{sign}\left(\Delta\hat{Q}_i\right) = \text{sign}\left(\Delta\hat{Q}_{i+2}\right) = \text{sign}\left(F^a_{i+\frac{1}{2}}\right) \qquad (8.19)$$

Thus, the corrected flux $F^c_{i+\frac{1}{2}}$ exists if and only if the anti-diffusive flux $F^a_{i+\frac{1}{2}}$ is aligned with (*i.e.*, has the same sign as) the gradient of $\hat{Q}$ as defined by the one-sided differences $\Delta\hat{Q}_i$ and $\Delta\hat{Q}_{i+2}$ on either side of the interface at $i+\frac{1}{2}$. Third, it may be shown (Exercise 8.9) that

$$F^c_{i+\frac{1}{2}} = 0 \quad \text{if} \quad \text{sign}(\Delta\hat{Q}_{i+2}) = -\text{sign}(\Delta\hat{Q}_i) \qquad (8.20)$$

This is a corollary to (8.19). If $\Delta\hat{Q}_i$ and $\Delta\hat{Q}_{i+2}$ have opposite signs, then it is not possible (in the context of FCT) to define a direction of the gradient of $\hat{Q}$ at $i+\frac{1}{2}$, and thus the corrected flux is set to zero. Fourth, in a more general sense, it may be shown that

$$
\begin{aligned}
F^c_{i+\frac{1}{2}} \geq 0 \quad &\text{if } \Delta\hat{Q}_{i+2} \geq 0 \text{ and } \Delta\hat{Q}_i \geq 0 \\
F^c_{i+\frac{1}{2}} \leq 0 \quad &\text{if } \Delta\hat{Q}_{i+2} \leq 0 \text{ and } \Delta\hat{Q}_i \leq 0 \\
F^c_{i+\frac{1}{2}} = 0 \quad &\text{if } \text{sign}(\Delta\hat{Q}_{i+2}) = -\text{sign}(\Delta\hat{Q}_i)
\end{aligned}
\qquad (8.21)
$$

This may be proven as follows. Consider $\Delta\hat{Q}_{i+2} \geq 0$ and $\Delta\hat{Q}_i \geq 0$. Assume

$S_{i+\frac{1}{2}} = 1$. Then $K_{i+\frac{1}{2}} \geq 0$ and $F^c_{i+\frac{1}{2}} \geq 0$. Assume $S_{i-\frac{1}{2}} = -1$. Then $K_{i+\frac{1}{2}} \leq 0$ and $F^c_{i+\frac{1}{2}} = 0$. The remaining three possible cases for the signs of $\Delta\hat{Q}_{i+2}$ and $\Delta\hat{Q}_i$ may be considered and the result proven.

From (8.20), $F^c_{i+\frac{1}{2}} = 0$ for Classes 6, 8, 9, and 11. Therefore, Q^{n+1}_i does not introduce any new extrema (*i.e.*, new maximum or minimum) relative to the set $(\hat{Q}_{i-2}, \ldots, \hat{Q}_{i+2})$ for these classes.

We now examine the remaining classes to ascertain whether new extrema can be introduced at i relative to the set $(\hat{Q}_{i-2}, \ldots, \hat{Q}_{i+2})$. Note that $(\hat{Q}_{i-2}, \ldots, \hat{Q}_{i+2})$ represents the minimum† set from the intermediate solution used to obtain Q^{n+1}_i. For reference, we combine (8.12) and (8.13),

$$Q^{n+1}_i = \hat{Q}_i - \frac{\Delta t}{\Delta x}[\underbrace{S_{i+\frac{1}{2}} \max\left[0, K_{i+\frac{1}{2}}\right]}_{F^c_{i+\frac{1}{2}}} - \underbrace{S_{i-\frac{1}{2}} \max\left[0, K_{i-\frac{1}{2}}\right]}_{F^c_{i-\frac{1}{2}}}] \qquad (8.22)$$

8.3.1.1 Class 1

For Class 1, the intermediate solution $\hat{Q}_i$ is monotonically increasing as shown in Fig. 8.8. There are four possible cases depending on the sign of $F^a_{i\pm\frac{1}{2}}$. We consider the possible generation of new extrema. For Class 1, the largest value of $\hat{Q}$ in the set $(\hat{Q}_{i-2}, \ldots, \hat{Q}_{i+2})$ is $\hat{Q}_{i+2}$, and the smallest value is $\hat{Q}_{i-2}$.

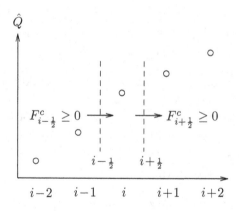

Fig. 8.8. Class 1

We first consider the possible creation of a new maximum at i. From

---

† Depending on the method of reconstruction used to obtain $F^a_{i\pm\frac{1}{2}}$, a larger set may be employed.

(8.22),

$$Q_i^{n+1} - \hat{Q}_{i+2} = \underbrace{-\Delta \hat{Q}_{i+1} + S_{i-\frac{1}{2}} \max\left[0, \frac{\Delta t}{\Delta x} K_{i-\frac{1}{2}}\right]}_{\Delta_1}$$

$$\underbrace{-\left\{\Delta \hat{Q}_{i+2} + S_{i+\frac{1}{2}} \max\left[0, \frac{\Delta t}{\Delta x} K_{i+\frac{1}{2}}\right]\right\}}_{\Delta_2} \quad (8.23)$$

We now show that the terms Δ_1 and Δ_2 are each nonpositive, thereby proving that no new maximum is created. For Δ_1, there are two possible cases. If $\text{sign}(F_{i-\frac{1}{2}}^a) = -1$, then $K_{i-\frac{1}{2}} < 0$ from (8.14) and therefore

$$\Delta_1 = -\Delta \hat{Q}_{i+1}$$

which is negative (Fig. 8.8). If $\text{sign}(F_{i-\frac{1}{2}}^a) = +1$, then

$$\Delta_1 = -\Delta \hat{Q}_{i+1} + \min\left[\frac{\Delta t}{\Delta x} F_{i-\frac{1}{2}}^a, \Delta \hat{Q}_{i+1}, \Delta \hat{Q}_{i-1}\right]$$

which is nonpositive since

$$-a + \min(a, b, \ldots) \leq 0 \quad (8.24)$$

for any scalar quantities a, b, $\ldots$. Thus, term Δ_1 is nonpositive.

For Δ_2, there are two possible cases. If $\text{sign}(F_{i+\frac{1}{2}}^a) = -1$, then $K_{i+\frac{1}{2}} < 0$ from (8.14) and therefore

$$\Delta_2 = -\Delta \hat{Q}_{i+2}$$

which is negative (Fig. 8.8). If $\text{sign}(F_{i+\frac{1}{2}}^a) = +1$, then

$$\Delta_2 = -\left\{\Delta \hat{Q}_{i+2} + \min\left[\frac{\Delta t}{\Delta x} F_{i+\frac{1}{2}}^a, \Delta \hat{Q}_{i+2}, \Delta \hat{Q}_i\right]\right\}$$

which is negative. Thus, the term Δ_2 is negative. Thus, no new maximum is created at i relative to the set $(\hat{Q}_{i-2}, \ldots, \hat{Q}_{i+2})$.

We next consider the possible creation of a new minimum at i. From (8.22),

$$Q_i^{n+1} - \hat{Q}_{i-2} = \underbrace{\Delta \hat{Q}_{i-1} + S_{i-\frac{1}{2}} \max\left[0, \frac{\Delta t}{\Delta x} K_{i-\frac{1}{2}}\right]}_{\Delta_1}$$

$$\underbrace{+\left\{\Delta \hat{Q}_i - S_{i+\frac{1}{2}} \max\left[0, \frac{\Delta t}{\Delta x} K_{i+\frac{1}{2}}\right]\right\}}_{\Delta_2} \quad (8.25)$$

We now show that the terms Δ_1 and Δ_2 are each nonnegative, thereby proving that no new minimum is created. For Δ_1, there are two possible cases. If $\text{sign}(F^a_{i-\frac{1}{2}}) = -1$, then $K_{i-\frac{1}{2}} < 0$ and therefore

$$\Delta_1 = \Delta\hat{Q}_{i-1}$$

which is positive (Fig. 8.8). If $\text{sign}(F^a_{i-\frac{1}{2}}) = +1$, then

$$\Delta_1 = \Delta\hat{Q}_{i-1} + \min\left[\frac{\Delta t}{\Delta x}F^a_{i-\frac{1}{2}}, \Delta\hat{Q}_{i+1}, \Delta\hat{Q}_{i-1}\right]$$

which is positive. Thus, Δ_1 is positive.

For Δ_2, there are two possible cases. If $\text{sign}(F^a_{i+\frac{1}{2}}) = -1$, then $K_{i+\frac{1}{2}} < 0$ and

$$\Delta_2 = \Delta\hat{Q}_i$$

which is positive (Fig. 8.8). If $\text{sign}(F^a_{i+\frac{1}{2}}) = +1$, then

$$\Delta_2 = \Delta\hat{Q}_i - \min\left[\frac{\Delta t}{\Delta x}F^a_{i+\frac{1}{2}}, \Delta\hat{Q}_{i+2}, \Delta\hat{Q}_i\right]$$

which is positive from (8.24). Thus, Δ_2 is positive. Therefore, no new minimum is created at i relative to the set $(\hat{Q}_{i-2}, \ldots, \hat{Q}_{i+2})$.

8.3.1.2 Class 2

For Class 2, the intermediate solution $\hat{Q}_i$ is monotonically increasing to $\hat{Q}_{i+1}$ and decreasing from $\hat{Q}_{i+1}$ to $\hat{Q}_{i+2}$ as shown in Fig. 8.9. Since $\text{sign}(\Delta\hat{Q}_{i+2}) = -\text{sign}(\Delta\hat{Q}_i)$, thus $F^c_{i+\frac{1}{2}} = 0$ from (8.21). Thus, from (8.22),

$$Q_i^{n+1} = \hat{Q}_i + \frac{\Delta t}{\Delta x}F^c_{i-\frac{1}{2}}$$

Since $\text{sign}(\Delta\hat{Q}_{i+1}) = \text{sign}(\Delta\hat{Q}_{i-1}) = +1$, therefore $F^c_{i-\frac{1}{2}} \geq 0$ from (8.21).

For Class 2, the largest value of $\hat{Q}$ in the set $(\hat{Q}_{i-2}, \ldots, \hat{Q}_{i+2})$ is $\hat{Q}_{i+1}$, and the smallest value is $\min(\hat{Q}_{i-2}, \hat{Q}_{i+2})$.

There are two possible cases depending on $\text{sign}(F^a_{i-\frac{1}{2}})$. If $\text{sign}(F^a_{i-\frac{1}{2}}) = -1$, $F^c_{i-\frac{1}{2}} = 0$ from (8.19) and thus $Q_i^{n+1} = \hat{Q}_i$, and hence no new extremum is created. If $\text{sign}(F^a_{i-\frac{1}{2}}) = +1$, we consider the possible creation of a new maximum. Now,

$$Q_i^{n+1} - \hat{Q}_{i+1} = -\Delta\hat{Q}_{i+1} + S_{i-\frac{1}{2}}\max\left[0, \frac{\Delta t}{\Delta x}K_{i-\frac{1}{2}}\right]$$

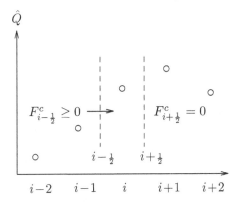

Fig. 8.9. Class 2

and thus

$$Q_i^{n+1} - \hat{Q}_{i+1} = -\Delta\hat{Q}_{i+1} + \min\left[\frac{\Delta t}{\Delta x}F_{i-\frac{1}{2}}^a, \Delta\hat{Q}_{i+1}, \Delta\hat{Q}_{i-1}\right]$$

which is nonpositive from (8.24). Thus, no new maximum is created relative to the set $(\hat{Q}_{i-2}, \ldots, \hat{Q}_{i+2})$.

We consider the possible creation of a new minimum when $\text{sign}(F_{i-\frac{1}{2}}^a) = +1$. The minimum value of $\hat{Q}$ in the set $(\hat{Q}_{i-2}, \ldots, \hat{Q}_{i+2})$ is $\min(\hat{Q}_{i-2}, \hat{Q}_{i+2})$. Now,

$$Q_i^{n+1} - \hat{Q}_{i-2} = \underbrace{\Delta\hat{Q}_i}_{\Delta_1} + \underbrace{\Delta\hat{Q}_{i-1} + S_{i-\frac{1}{2}}\max\left[0, \frac{\Delta t}{\Delta x}K_{i-\frac{1}{2}}\right]}_{\Delta_2}$$

Now $\Delta_1 > 0$ and

$$\Delta_2 = \Delta\hat{Q}_{i-1} + \min\left[\frac{\Delta t}{\Delta x}F_{i-\frac{1}{2}}^a, \Delta\hat{Q}_{i+1}, \Delta\hat{Q}_{i-1}\right]$$

and thus $\Delta_2 > 0$. Now,

$$Q_i^{n+1} - \min(\hat{Q}_{i-2}, \hat{Q}_{i+2}) \geq Q_i^{n+1} - \hat{Q}_{i-2}$$

and thus no new minimum is created.

8.3.1.3 Class 3

For Class 3, the intermediate solution $\hat{Q}_i$ is monotonically increasing to $\hat{Q}_i$, decreasing from $\hat{Q}_i$ to $\hat{Q}_{i+1}$, and increasing to $\hat{Q}_{i+2}$ as shown in Fig. 8.10.

Since $\text{sign}(\Delta\hat{Q}_{i+1}) = -\text{sign}(\Delta\hat{Q}_{i-1})$, thus $F^c_{i-\frac{1}{2}} = 0$. Thus,

$$Q_i^{n+1} = \hat{Q}_i - \frac{\Delta t}{\Delta x}F^c_{i+\frac{1}{2}}$$

Since $\text{sign}(\Delta\hat{Q}_{i+2}) = \text{sign}(\Delta\hat{Q}_i) = +1$, thus $F^c_{i+\frac{1}{2}} \geq 0$. For Class 3, the largest value of $\hat{Q}$ in the set $(\hat{Q}_{i-2}, \ldots, \hat{Q}_{i+2})$ is $\max(\hat{Q}_i, \hat{Q}_{i+2})$, and the smallest value is $\min(\hat{Q}_{i-2}, \hat{Q}_{i+1})$.

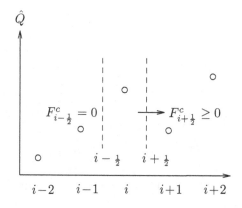

Fig. 8.10. Class 3

There are two possible cases depending on $\text{sign}(F^a_{i+\frac{1}{2}})$. If $\text{sign}(F^a_{i+\frac{1}{2}}) = -1$, then $F^c_{i+\frac{1}{2}} = 0$ from (8.19) and no new extremum is created. If $\text{sign}(F^a_{i+\frac{1}{2}}) = +1$, we examine the possible creation of a new extremum. Now,

$$Q_i^{n+1} - \hat{Q}_i = -\underbrace{\min\left[\frac{\Delta t}{\Delta x}F^a_{i+\frac{1}{2}}, \Delta\hat{Q}_{i+2}, \Delta\hat{Q}_i\right]}_{\Delta_1}$$

where $\Delta_1 > 0$. Thus,

$$Q_i^{n+1} - \max(\hat{Q}_i, \hat{Q}_{i+2}) \leq Q_i^{n+1} - \hat{Q}_i < 0$$

Thus, no new maximum is formed. Also,

$$Q_i^{n+1} - \hat{Q}_{i-1} = \Delta\hat{Q}_i - \underbrace{\min\left[\frac{\Delta t}{\Delta x}F^a_{i+\frac{1}{2}}, \Delta\hat{Q}_{i+2}, \Delta\hat{Q}_i\right]}_{\Delta_1}$$

and from (8.24), $\Delta_1 \geq 0$. Now,

$$Q_i^{n+1} - \min(\hat{Q}_{i-2}, \hat{Q}_{i+1}) = \underbrace{Q_i^{n+1} - \hat{Q}_{i-1}}_{\geq 0} + \underbrace{\hat{Q}_{i-1} - \min(\hat{Q}_{i-2}, \hat{Q}_{i+1})}_{\geq 0}$$

Thus, no new minimum is formed.

8.3.1.4 Class 4

For Class 4, the intermediate solution $\hat{Q}_i$ is monotonically increasing to $\hat{Q}_{i-1}$, decreasing from $\hat{Q}_{i-1}$ to $\hat{Q}_i$, and increasing to $\hat{Q}_{i+2}$ as shown in Fig. 8.11. Since $\text{sign}(\Delta\hat{Q}_{i+2}) = -\text{sign}(\Delta\hat{Q}_i)$, thus $F^c_{i+\frac{1}{2}} = 0$. Since $\text{sign}(\Delta\hat{Q}_{i+1}) = \text{sign}(\Delta\hat{Q}_{i-1}) = +1$, thus $F^c_{i-\frac{1}{2}} \geq 0$. Thus,

$$Q_i^{n+1} = \hat{Q}_i + \frac{\Delta t}{\Delta x} F^c_{i-\frac{1}{2}}$$

For Class 4, the largest value of $\hat{Q}$ in the set $(\hat{Q}_{i-2}, \ldots, \hat{Q}_{i+2})$ is $\max(\hat{Q}_{i-1}, \hat{Q}_{i+2})$, and the smallest value is $\min(\hat{Q}_{i-2}, \hat{Q}_i)$.

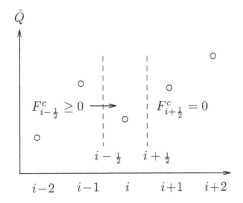

Fig. 8.11. Class 4

There are two cases depending on $\text{sign}(F^a_{i-\frac{1}{2}})$. If $\text{sign}(F^a_{i-\frac{1}{2}}) = -1$, then $F^c_{i-\frac{1}{2}} = 0$ and no new extremum is created. If $\text{sign}(F^a_{i-\frac{1}{2}}) = +1$, we examine the possible creation of a new extremum. Now,

$$Q_i^{n+1} - \hat{Q}_{i+1} = \underbrace{-\Delta\hat{Q}_{i+1} + \min\left[\frac{\Delta t}{\Delta x}F^a_{i-\frac{1}{2}}, \Delta\hat{Q}_{i+1}, \Delta\hat{Q}_{i-1}\right]}_{\Delta_1}$$

where from (8.24), $\Delta_1 \leq 0$. Now,

$$Q_i^{n+1} - \max(\hat{Q}_{i-1}, \hat{Q}_{i+2}) = \underbrace{Q_i^{n+1} - \hat{Q}_{i+1}}_{\leq 0} + \underbrace{\hat{Q}_{i+1} - \max(\hat{Q}_{i-1}, \hat{Q}_{i+2})}_{\leq 0}$$

and thus no new maximum is created. Furthermore,

$$Q_i^{n+1} - \hat{Q}_i = \underbrace{\min\left[\frac{\Delta t}{\Delta x}F^a_{i-\frac{1}{2}}, \Delta\hat{Q}_{i+1}, \Delta\hat{Q}_{i-1}\right]}_{\Delta_2}$$

where $\Delta_2 \geq 0$. Now,

$$Q_i^{n+1} - \min(\hat{Q}_{i-2}, \hat{Q}_i) = \underbrace{Q_i^{n+1} - \hat{Q}_i}_{\geq 0} + \underbrace{\hat{Q}_i - \min(\hat{Q}_{i-2}, \hat{Q}_i)}_{\geq 0}$$

and thus no new minimum is formed.

8.3.1.5 Class 5

For Class 5, the intermediate solution $\hat{Q}_i$ is monotonically decreasing to $\hat{Q}_{i-1}$ and increasing to $\hat{Q}_{i+2}$ as shown in Fig. 8.12. Since $\text{sign}(\Delta\hat{Q}_{i+1}) = -\text{sign}(\Delta\hat{Q}_{i-1})$, thus $F_{i-\frac{1}{2}}^c = 0$. Thus,

$$Q_i^{n+1} = \hat{Q}_i - \frac{\Delta t}{\Delta x} F_{i+\frac{1}{2}}^c$$

Since $\text{sign}(\Delta\hat{Q}_{i+2}) = \text{sign}(\Delta\hat{Q}_i) = +1$, thus $F_{i+\frac{1}{2}}^c \geq 0$. For Class 5, the largest value of $\hat{Q}$ in the set $(\hat{Q}_{i-2}, \ldots, \hat{Q}_{i+2})$ is $\max(\hat{Q}_{i-2}, \hat{Q}_{i+2})$, and the smallest value is $\hat{Q}_{i-1}$.

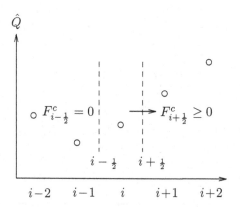

Fig. 8.12. Class 5

There are two possible cases depending on $\text{sign}(F_{i+\frac{1}{2}}^a)$. If $\text{sign}(F_{i+\frac{1}{2}}^a) = -1$, then $F_{i+\frac{1}{2}}^c = 0$ and no new extremum is created. If $\text{sign}(F_{i+\frac{1}{2}}^a) = +1$, we examine the possibility of a new extremum. Now,

$$Q_i^{n+1} - \hat{Q}_i = -\underbrace{\min\left[\frac{\Delta t}{\Delta x} F_{i+\frac{1}{2}}^a, \Delta\hat{Q}_{i+2}, \Delta\hat{Q}_i\right]}_{\Delta_1}$$

where $\Delta_1 \geq 0$. Now,

$$Q_i^{n+1} - \max(\hat{Q}_{i-2}, \hat{Q}_{i+2}) = \underbrace{Q_i^{n+1} - \hat{Q}_i}_{\leq 0} + \underbrace{\hat{Q}_i - \max(\hat{Q}_{i-2}, \hat{Q}_{i+2})}_{\leq 0}$$

Thus no new maximum is formed. Furthermore,

$$Q_i^{n+1} - \hat{Q}_{i-1} = \underbrace{\Delta\hat{Q}_i - \min\left[\frac{\Delta t}{\Delta x}F_{i+\frac{1}{2}}^a, \Delta\hat{Q}_{i+2}, \Delta\hat{Q}_i\right]}_{\Delta_2}$$

where from (8.24), $\Delta_2 \geq 0$. Thus no new minimum is formed.

8.3.1.6 Class 6

For Class 6, the intermediate solution $\hat{Q}_i$ is monotonically increasing to $\hat{Q}_i$ and decreasing to $\hat{Q}_{i+2}$ as shown in Fig. 8.13. Since $\text{sign}(\Delta\hat{Q}_{i+1}) = -\text{sign}(\Delta\hat{Q}_{i-1})$, thus $F_{i-\frac{1}{2}}^c = 0$. Since $\text{sign}(\Delta\hat{Q}_{i+2}) = -\text{sign}(\Delta\hat{Q}_i)$, thus $F_{i+\frac{1}{2}}^c = 0$. Thus, $Q_i^{n+1} = \hat{Q}_i$, and no new maximum or minimum in formed.

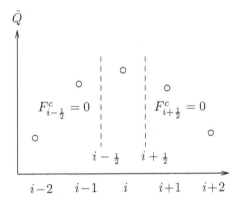

Fig. 8.13. Class 6

8.3.1.7 Class 7

For Class 7, the intermediate solution $\hat{Q}_i$ is oscillatory as shown in Fig. 8.14. Since $\text{sign}(\Delta\hat{Q}_{i+1}) = \text{sign}(\Delta\hat{Q}_{i-1}) = +1$, thus $F_{i-\frac{1}{2}}^c \geq 0$. Since $\text{sign}(\Delta\hat{Q}_{i+2}) = \text{sign}(\Delta\hat{Q}_i) = -1$, thus $F_{i+\frac{1}{2}}^c \leq 0$.

From (8.22),

$$Q_i^{n+1} = \hat{Q}_i - \frac{\Delta t}{\Delta x}\left\{S_{i+\frac{1}{2}}\max\left[0, K_{i+\frac{1}{2}}\right] - S_{i-\frac{1}{2}}\max\left[0, K_{i-\frac{1}{2}}\right]\right\}$$

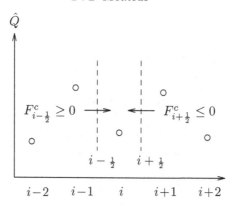

Fig. 8.14. Class 7

Assume that the anti-diffusive fluxes $F^a_{i\pm\frac{1}{2}}$ are aligned with the local gradients, *i.e.*, $S_{i-\frac{1}{2}} = +1$ and $S_{i+\frac{1}{2}} = -1$. We show by example that a new maximum can be formed. Assume

$$\Delta\hat{Q}_{i-1} = 5, \ \Delta\hat{Q}_i = -1, \ \Delta\hat{Q}_{i+1} = 3, \ \Delta\hat{Q}_{i+2} = -2, \ \frac{\Delta t}{\Delta x}F^a_{i-\frac{1}{2}} > 3, \ \frac{\Delta t}{\Delta x}\left|F^a_{i-\frac{1}{2}}\right| > 1$$

Then

$$Q^{n+1}_1 = \hat{Q}_i + 4$$

which is a new maximum relative to the set $(\hat{Q}_{i-2}, \ldots, \hat{Q}_{i+2})$. A new minimum cannot be formed, however, since

$$Q^{n+1}_i - \hat{Q}_i = \underbrace{-S_{i+\frac{1}{2}} \max\left[0, \frac{\Delta t}{\Delta x}K_{i+\frac{1}{2}}\right]}_{\Delta_1} + \underbrace{S_{i-\frac{1}{2}} \max\left[0, \frac{\Delta t}{\Delta x}K_{i-\frac{1}{2}}\right]}_{\Delta_2}$$

and both Δ_1 and Δ_2 are nonnegative.

8.3.1.8 Class 8

For Class 8, the intermediate solution $\hat{Q}_i$ is monotonically decreasing to $\hat{Q}_{i-1}$, increasing to $\hat{Q}_{i+1}$, and decreasing to $\hat{Q}_{i+2}$ as shown in Fig. 8.15. Since $\text{sign}(\Delta\hat{Q}_{i+1}) = -\text{sign}(\Delta\hat{Q}_{i-1})$, thus $F^c_{i-\frac{1}{2}} = 0$. Similarly, since $\text{sign}(\Delta\hat{Q}_{i+2}) = -\text{sign}(\Delta\hat{Q}_i)$, thus $F^c_{i+\frac{1}{2}} = 0$. Thus, $Q^{n+1} = \hat{Q}_i$ and no new maximum or minimum is formed.

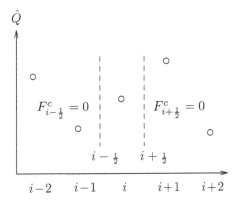

Fig. 8.15. Class 8

8.3.1.9 Class 9

For Class 9, the intermediate solution $\hat{Q}_i$ is monotonically increasing to $\hat{Q}_{i-1}$, decreasing to $\hat{Q}_{i+1}$, and increasing to $\hat{Q}_{i+2}$ as shown in Fig. 8.16. This is equivalent to the reflection of Class 8 about i. Since $\text{sign}(\Delta\hat{Q}_{i+1}) = -\text{sign}(\Delta\hat{Q}_{i-1})$, thus $F^c_{i-\frac{1}{2}} = 0$. Similarly, since $\text{sign}(\Delta\hat{Q}_{i+2}) = -\text{sign}(\Delta\hat{Q}_i)$, thus $F^c_{i+\frac{1}{2}} = 0$. Thus, $Q^{n+1} = \hat{Q}_i$ and no new maximum or minimum is formed.

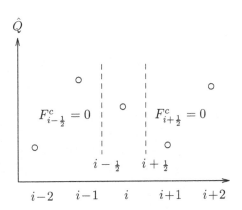

Fig. 8.16. Class 9

8.3.1.10 Class 10

For Class 10, the intermediate solution $\hat{Q}_i$ is oscillatory. Since $\text{sign}(\Delta\hat{Q}_{i+1}) = \text{sign}(\Delta\hat{Q}_{i-1}) = -1$, thus $F^c_{i-\frac{1}{2}} \leq 0$. Similarly, since $\text{sign}(\Delta\hat{Q}_{i+2}) = \text{sign}(\Delta\hat{Q}_i) = +1$, thus $F^c_{i+\frac{1}{2}} \geq 0$.

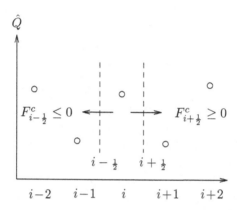

Fig. 8.17. Class 10

From (8.22),

$$Q_i^{n+1} = \hat{Q}_i - \frac{\Delta t}{\Delta x}\left\{S_{i+\frac{1}{2}}\max\left[0, K_{i+\frac{1}{2}}\right] - S_{i-\frac{1}{2}}\max\left[0, K_{i-\frac{1}{2}}\right]\right\}$$

Assume that the anti-diffusive fluxes $F^a_{i\pm\frac{1}{2}}$ are aligned with the local gradients, *i.e.*, $S_{i-\frac{1}{2}} = -1$ and $S_{i+\frac{1}{2}} = +1$. We show by example that a new minimum can be formed. Assume

$$\Delta\hat{Q}_{i-1} = -2, \ \Delta\hat{Q}_i = 1, \ \Delta\hat{Q}_{i+1} = -2, \ \Delta\hat{Q}_{i+2} = 2, \ \frac{\Delta t}{\Delta x}\left|F^a_{i-\frac{1}{2}}\right| > 2, \ \frac{\Delta t}{\Delta x}F^a_{i-\frac{1}{2}} > 1$$

Then

$$Q_i^{n+1} = \hat{Q}_i - 3$$

which is a new minimum relative to the set $(\hat{Q}_{i-2}, \ldots, \hat{Q}_{i+2})$. A new maximum cannot be formed, however, since

$$Q_i^{n+1} - \hat{Q}_i = \underbrace{-S_{i+\frac{1}{2}}\max\left[0, \frac{\Delta t}{\Delta x}K_{i+\frac{1}{2}}\right]}_{\Delta_1} + \underbrace{S_{i-\frac{1}{2}}\max\left[0, \frac{\Delta t}{\Delta x}K_{i-\frac{1}{2}}\right]}_{\Delta_2}$$

and both Δ_1 and Δ_2 are nonpositive.

8.3.1.11 Class 11

For Class 11, the intermediate solution $\hat{Q}_i$ is monotonically decreasing to $\hat{Q}_i$ and increasing to $\hat{Q}_{i+2}$ as shown in Fig. 8.18. Since $\text{sign}(\Delta\hat{Q}_{i+1}) = -\text{sign}(\Delta\hat{Q}_{i-1})$, thus $F^c_{i-\frac{1}{2}} = 0$. Similarly, since $\text{sign}(\Delta\hat{Q}_{i+2}) = -\text{sign}(\Delta\hat{Q}_i)$, thus $F^c_{i+\frac{1}{2}} = 0$. Thus, $Q^{n+1} = \hat{Q}_i$ and no new maximum or minimum is formed.

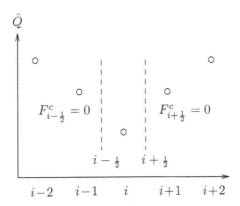

Fig. 8.18. Class 11

8.3.1.12 Class 12

For Class 12, the intermediate solution $\hat{Q}_i$ is increasing to $\hat{Q}_{i-1}$ and decreasing to $\hat{Q}_{i+2}$ as shown in Fig. 8.19. Since $\text{sign}(\Delta\hat{Q}_{i+1}) = -\text{sign}(\Delta\hat{Q}_{i-1})$, thus $F^c_{i-\frac{1}{2}} = 0$. Thus,

$$Q_i^{n+1} = \hat{Q}_i - \frac{\Delta t}{\Delta x} F^c_{i+\frac{1}{2}}$$

Since $\text{sign}(\Delta\hat{Q}_{i+2}) = \text{sign}(\Delta\hat{Q}_i) = -1$, thus $F^c_{i+\frac{1}{2}} \leq 0$.

There are two possible cases depending on $\text{sign}(F^a_{i+\frac{1}{2}})$. If $\text{sign}(F^a_{i+\frac{1}{2}}) = +1$, then $F^c_{i+\frac{1}{2}} = 0$ and no new extremum is created. If $\text{sign}(F^a_{i+\frac{1}{2}}) = -1$, we examine the possibility of a new extremum. Now,

$$Q_i^{n+1} - \hat{Q}_{i-1} = \underbrace{\Delta\hat{Q}_i}_{\Delta_1} + \underbrace{\min\left[\frac{\Delta t}{\Delta x}\left|F^a_{i+\frac{1}{2}}\right|, -\Delta\hat{Q}_{i+2}, -\Delta\hat{Q}_i\right]}_{\Delta_2}$$

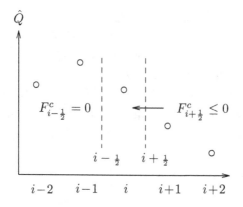

Fig. 8.19. Class 12

Since $\Delta_2 \leq -\Delta_1$, no new maximum is formed. Furthermore,

$$Q_i^{n+1} = \hat{Q}_i + \underbrace{\min\left[\frac{\Delta t}{\Delta x}F_{i+\frac{1}{2}}^a, -\Delta\hat{Q}_{i+2}, -\Delta\hat{Q}_i\right]}_{\Delta_1}$$

Since $\Delta_1 \geq 0$, no new minimum is formed.

8.3.1.13 Class 13

For Class 13, the intermediate solution $\hat{Q}_i$ is decreasing to $\hat{Q}_{i-1}$, increasing to $\hat{Q}_i$, and decreasing to $\hat{Q}_{i+2}$ as shown in Fig. 8.20. Since $\mathrm{sign}(\Delta\hat{Q}_{i+2}) = -\mathrm{sign}(\Delta\hat{Q}_i)$, thus $F_{i+\frac{1}{2}}^c = 0$. Thus,

$$Q_i^{n+1} = \hat{Q}_i + \frac{\Delta t}{\Delta x}F_{i-\frac{1}{2}}^c$$

Since $\mathrm{sign}(\Delta\hat{Q}_{i+1}) = \mathrm{sign}(\Delta\hat{Q}_{i-1}) = -1$, thus $F_{i-\frac{1}{2}}^c \leq 0$.

There are two possible cases depending on $\mathrm{sign}(F_{i-\frac{1}{2}}^a)$. If $\mathrm{sign}(F_{i-\frac{1}{2}}^a) = +1$, then $F_{i-\frac{1}{2}}^c = 0$ and no new extremum is created. If $\mathrm{sign}(F_{i-\frac{1}{2}}^a) = -1$, we examine the possibility of a new extremum. Now,

$$Q_i^{n+1} - \hat{Q}_i = -\underbrace{\min\left[\frac{\Delta t}{\Delta x}\left|F_{i+\frac{1}{2}}^a\right|, -\Delta\hat{Q}_{i+1}, -\Delta\hat{Q}_{i-1}\right]}_{\Delta_1}$$

Since $\Delta_1 \geq 0$, no new maximum is formed. Furthermore,

$$Q_i^{n+1} - \hat{Q}_{i+1} = \underbrace{-\Delta\hat{Q}_{i+1}}_{\Delta_1} - \underbrace{\min\left[\frac{\Delta t}{\Delta x}\left|F_{i-\frac{1}{2}}^a\right|, -\Delta\hat{Q}_{i+1}, -\Delta\hat{Q}_{i-1}\right]}_{\Delta_2}$$

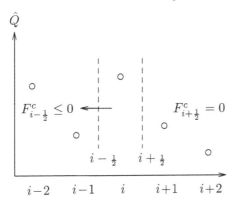

Fig. 8.20. Class 13

Since $\Delta_2 \leq \Delta_1$, no new minimum is formed.

8.3.1.14 Class 14

For Class 14, the intermediate solution $\hat{Q}_i$ is decreasing to $\hat{Q}_i$, increasing to $\hat{Q}_{i+1}$, and decreasing to $\hat{Q}_{i+2}$ as shown in Fig. 8.21. Since $\text{sign}(\Delta\hat{Q}_{i+1}) = -\text{sign}(\Delta\hat{Q}_{i-1})$, thus $F^c_{i-\frac{1}{2}} = 0$. Thus,

$$Q_i^{n+1} = \hat{Q}_i - \frac{\Delta t}{\Delta x}F^c_{i+\frac{1}{2}}$$

Since $\text{sign}(\Delta\hat{Q}_{i+2}) = \text{sign}(\Delta\hat{Q}_i) = -1$, thus $F^c_{i+\frac{1}{2}} \leq 0$.

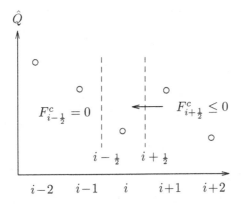

Fig. 8.21. Class 14

There are two possible cases depending on $\text{sign}(F^a_{i+\frac{1}{2}})$. If $\text{sign}(F^a_{i+\frac{1}{2}}) = +1$, then $F^c_{i+\frac{1}{2}} = 0$ and no new extremum is created. If $\text{sign}(F^a_{i+\frac{1}{2}}) = -1$, we examine the possibility of a new extremum. Now,

$$Q_i^{n+1} - \hat{Q}_i = \underbrace{\min\left[\frac{\Delta t}{\Delta x}\left|F_{i+\frac{1}{2}}^a\right|, -\Delta\hat{Q}_{i+2}, -\Delta\hat{Q}_i\right]}_{\Delta_1}$$

Since $\Delta_1 \geq 0$, no new minimum is formed. Furthermore,

$$Q_i^{n+1} - \hat{Q}_{i-1} = \underbrace{\Delta\hat{Q}_i}_{\Delta_1} + \underbrace{\min\left[\frac{\Delta t}{\Delta x}\left|F_{i+\frac{1}{2}}^a\right|, -\Delta\hat{Q}_{i+2}, -\Delta\hat{Q}_i\right]}_{\Delta_2}$$

Since $\Delta_2 \leq \Delta_1$, no new maximum is formed.

8.3.1.15 Class 15

For Class 15, the intermediate solution $\hat{Q}_i$ is decreasing to $\hat{Q}_{i+1}$ and increasing to $\hat{Q}_{i+2}$ as shown in Fig. 8.22. Since $\text{sign}(\Delta\hat{Q}_{i+1}) = -\text{sign}(\Delta\hat{Q}_i)$, thus $F_{i+\frac{1}{2}}^c = 0$. Thus,

$$Q_i^{n+1} = \hat{Q}_i - \frac{\Delta t}{\Delta x}F_{i-\frac{1}{2}}^c$$

Since $\text{sign}(\Delta\hat{Q}_{i+1}) = \text{sign}(\Delta\hat{Q}_{i-1}) = -1$, thus $F_{i-\frac{1}{2}}^c \leq 0$.

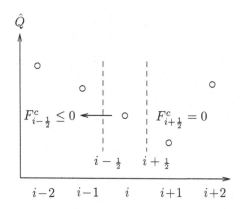

Fig. 8.22. Class 15

There are two possible cases depending on $\text{sign}(F_{i-\frac{1}{2}}^a)$. If $\text{sign}(F_{i-\frac{1}{2}}^a) = +1$, then $F_{i-\frac{1}{2}}^c = 0$ and no new extremum is created. If $\text{sign}(F_{i-\frac{1}{2}}^a) = -1$, we examine the possibility of a new extremum. Now,

$$Q_i^{n+1} - \hat{Q}_i = -\underbrace{\min\left[\frac{\Delta t}{\Delta x}\left|F_{i-\frac{1}{2}}^a\right|, -\Delta\hat{Q}_{i+1}, -\Delta\hat{Q}_{i-1}\right]}_{\Delta_1}$$

Since $\Delta_1 \geq 0$, no new maximum is formed. Furthermore,

$$Q_i^{n+1} - \hat{Q}_{i+1} = \underbrace{-\Delta\hat{Q}_{i+1}}_{\Delta_1} - \underbrace{\min\left[\frac{\Delta t}{\Delta x}\left|F_{i-\frac{1}{2}}^a\right|, -\Delta\hat{Q}_{i+1}, -\Delta\hat{Q}_{i-1}\right]}_{\Delta_2}$$

Since $\Delta_2 \leq \Delta_1$, no new minimum is formed.

8.3.1.16 Class 16

For Class 16, the intermediate solution $\hat{Q}_i$ is monotonically decreasing as shown in Fig. 8.23. There are four possible cases depending on the sign of $F_{i\pm\frac{1}{2}}^a$. This case is similar to Class 1. No new extrema are created relative to the set $(\hat{Q}_{i-2}, \ldots, \hat{Q}_{i+2})$. The proof is left as an exercise (Exercise 8.10).

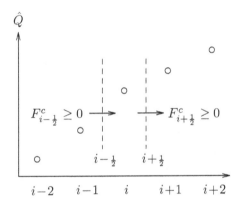

Fig. 8.23. Class 16

8.3.2 Accuracy, Consistency, and Convergence

We consider the problem described in Section 2.8. The initial condition is defined by (2.92) with $\epsilon = 0.1$ and the domain is $0 < \kappa x < 2\pi$. The low-order flux $F_{i\pm\frac{1}{2}}^l$ is computed using a first-order reconstruction ($Q_{i+\frac{1}{2}}^l = Q_i$ and $Q_{i-\frac{1}{2}}^r = Q_i$), and the high-order flux is computed using a third-order upwind biased reconstruction (4.19) with $\kappa = \frac{1}{3}$. The flux algorithm is Godunov's Method. The norm, defined by (3.102), is evaluated at $\kappa a_o t = 7$, which is prior to the shock formation. The timestep Δt is determined by

$$\Delta t = \mathcal{C}\Delta t_{CFL} \tag{8.26}$$

where the Courant number $\mathcal{C} = 0.46$ and Δt_{CFL} is calculated according to (5.8) using the initial condition. The timestep Δt is held in fixed ratio to the grid spacing Δx, and therefore Δt and Δx are decreased by the same ratio. The exact solution $Q_i^{n,e}$ is obtained from (2.85) to (2.88).

The convergence is displayed in Fig. 8.24. The solution converges linearly† to the exact solution for sufficiently small Δt. The linear convergence is a consequence of the first-order temporal integration. The computed result (using 100 cells) and exact solution are displayed in Fig. 8.25 for $\kappa a_o t = 7$ and $\mathcal{C} = 0.45$. The error in amplitude is associated with the first-order time integration and is significantly reduced by decreasing the timestep (Fig. 8.26).

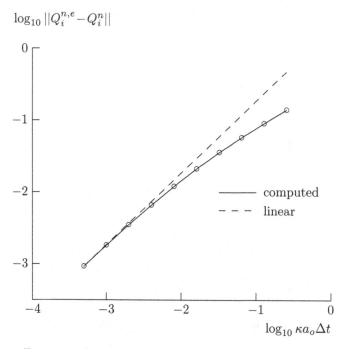

Fig. 8.24. Convergence for Flux Corrected Transport

† The line in Fig. 8.24 is

$$\|Q_i^{n,e} - Q_i^n\|_{\kappa a_o \Delta t} = \left(\frac{\kappa a_o \Delta t}{5 \times 10^{-4}} \right) \|Q_i^{n,e} - Q_i^n\|_{\kappa a_o \Delta t = 5 \times 10^{-4}}$$

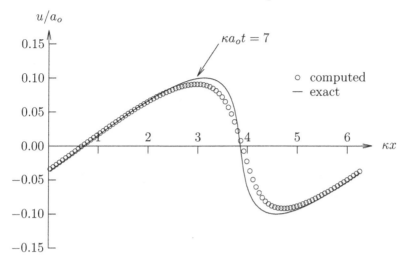

Fig. 8.25. Computed and exact solutions ($\kappa a_o \Delta t = 0.025$)

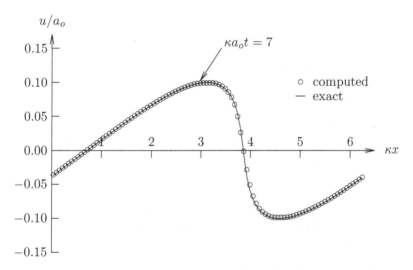

Fig. 8.26. Computed and exact solutions ($\kappa a_o \Delta t = 0.00125$)

8.3.3 Total Variation

We examine the Total Variation of the solution of the problem described in Section 2.8 with $\epsilon = 0.1$ and the domain $0 < \kappa x < 2\pi$ is discretized into 100 uniform cells where $\kappa = 2\pi L^{-1}$. Periodic boundary conditions are imposed on the left and right boundaries. The timestep is defined by (5.8) with $\mathcal{C} = 0.46$, and a first-order time integration is employed. The Total Variation of ρu is shown in Fig. 8.27 for four different high-order flux algorithms (Godunov, Roe, Osher, and Steger-Warming). The low-order flux $F^l_{i\pm\frac{1}{2}}$ is computed using a first-order reconstruction ($Q^l_{i+\frac{1}{2}} = Q_i$ and $Q^r_{i-\frac{1}{2}} = Q_i$), and the high-order flux is computed using a third-order upwind-biased reconstruction (4.19) with $\kappa = \frac{1}{3}$. All flux algorithms are evidently TVD with the incorporation of FCT. The decrease in TV is associated with the magnitude of Δt. In Fig. 8.28, the TV is shown for a second-order Runge-Kutta time integration with the same Δt. The decrease in TV is virtually the same as in Fig. 8.27. A reduction in Δt to $\Delta t = 0.00125$ virtually eliminates the drop in TV as shown in Fig. 8.29. The velocity profile is displayed at $\kappa a_o t = 14$ in Figs. 8.30 and 8.31 for $\Delta t = 0.025$ using first-order Euler and second-order Runge-Kutta integrations, respectively. The shock wave is diffused over several cells. A reduction in Δt to $\Delta t = 0.00125$ minimizes the diffusion of the shock as shown in Fig. 8.32.

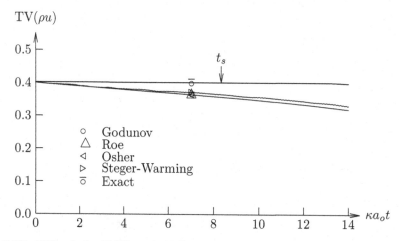

Fig. 8.27. TV(ρu) for FCT with different flux algorithms (first-order Euler with $\kappa a_o \Delta t = 0.025$)

8.4 MUSCL-Hancock Method

The MUSCL-Hancock method (Quirk, 1994; VanLeer, 1985) is a TVD-like algorithm based on the concept of a slope limiter. It is strictly TVD for a linear equation; however, experience has shown that it achieves TVD or near-TVD performance for the Euler equations (Toro, 1997). We present here one version of the MUSCL-Hancock method.

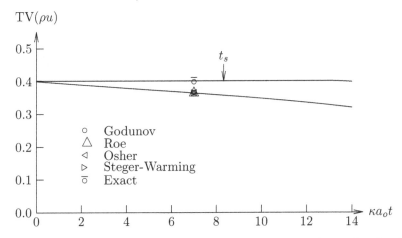

Fig. 8.28. TV(ρu) for FCT with different flux algorithms (second-order Runge-Kutta with $\Delta t = 0.025$)

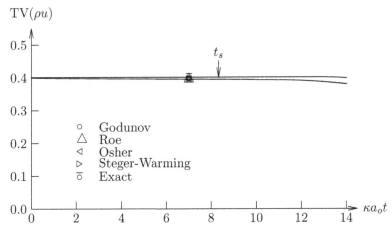

Fig. 8.29. TV(ρu) for FCT with different flux algorithms (first-order Euler with $\Delta t = 0.00125$)

8.4.1 Algorithm

Consider the semi-discrete form (3.9) of the Euler equations

$$\frac{dQ_i}{dt} + \frac{\left(F_{i+\frac{1}{2}} - F_{i-\frac{1}{2}}\right)}{\Delta x} = 0 \tag{8.27}$$

The MUSCL-Hancock Method proceeds in three steps. First, the flow variables are reconstructed within cell i using a linear expression:

$$Q(x) = Q_i + \frac{\xi}{\Delta x}\hat{\Delta}_i \tag{8.28}$$

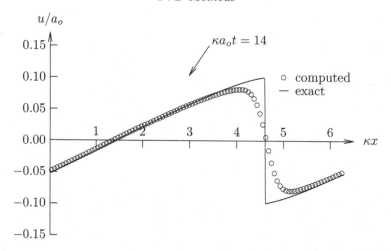

Fig. 8.30. Velocity at $\kappa a_o t = 14$ for FCT with Godunov's Method and third-order reconstruction (first-order Euler with $\Delta t = 0.025$)

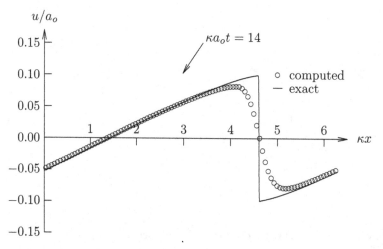

Fig. 8.31. Velocity at $\kappa a_o t = 14$ for FCT with Godunov's Method and third-order reconstruction (second-order Runge-Kutta with $\Delta t = 0.025$)

where

$$\xi = x - x_i \tag{8.29}$$

and $\hat{\Delta}_i$ is a limited linear combination of $\Delta Q_{i-\frac{1}{2}}$ and $\Delta Q_{i+\frac{1}{2}}$ according to

$$\hat{\Delta}_i = \varsigma(\varrho)\Delta_i \tag{8.30}$$

where Δ_i is

$$\Delta_i = \tfrac{1}{2}(1 - \kappa)\Delta Q_{i-\frac{1}{2}} + \tfrac{1}{2}(1 + \kappa)\Delta Q_{i+\frac{1}{2}} \tag{8.31}$$

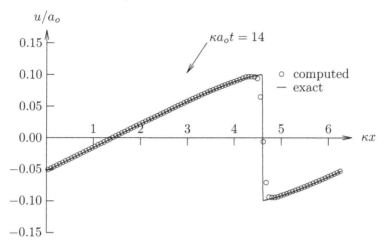

Fig. 8.32. Velocity at $\kappa a_o t = 14$ for FCT with Godunov's Method and third-order reconstruction (second-order Runge-Kutta with $\Delta t = 0.00125$)

and $-1 \leq \kappa \leq 1$ and $\varsigma(\varrho)$ is a slope limiter that depends on the ratio†

$$\varrho = \frac{\Delta Q_{i-\frac{1}{2}}}{\Delta Q_{i+\frac{1}{2}}} \tag{8.32}$$

A slope limiter analogous to the MinMod flux limiter is (Toro, 1997)

$$\varsigma = \begin{cases} 0 & \varrho \leq 0 \\ \varrho & 0 < \varrho \leq 1 \\ \min(1, \hat{\varsigma}(\varrho)) & \varrho > 1 \end{cases} \tag{8.33}$$

where

$$\hat{\varsigma} = \frac{2}{1 - \kappa + (1 + \kappa)\varrho} \tag{8.34}$$

Thus,

$$\begin{aligned} Q^l_{i+\frac{1}{2}} &= Q_i + \tfrac{1}{2}\hat{\Delta}_i \\ Q^r_{i-\frac{1}{2}} &= Q_i - \tfrac{1}{2}\hat{\Delta}_i \end{aligned} \tag{8.35}$$

Second, the boundary values are updated in time by $\frac{1}{2}\Delta t$ according to

$$\begin{aligned} \hat{Q}^l_{i+\frac{1}{2}} &= Q^l_{i+\frac{1}{2}} - \frac{1}{2}\frac{\Delta t}{\Delta x}\left(F_{i+\frac{1}{2}} - F_{i-\frac{1}{2}}\right) \\ \hat{Q}^r_{i-\frac{1}{2}} &= Q^r_{i-\frac{1}{2}} - \frac{1}{2}\frac{\Delta t}{\Delta x}\left(F_{i+\frac{1}{2}} - F_{i-\frac{1}{2}}\right) \end{aligned} \tag{8.36}$$

† The ratio in (8.32) is evaluated for each of the conservative variables. Thus, for the conservation of mass, $\varrho = \Delta\rho_{i-\frac{1}{2}}/\Delta\rho_{i+\frac{1}{2}}$.

where $F_{i+\frac{1}{2}}$ and $F_{i-\frac{1}{2}}$ are computed from (2.3) using $Q^l_{i+\frac{1}{2}}$ and $Q^r_{i-\frac{1}{2}}$, respectively. Third, the flow variables are updated according to

$$Q_i^{n+1} = Q_i^n - \frac{\Delta t}{\Delta x}\left(\hat{F}_{i+\frac{1}{2}} - \hat{F}_{i-\frac{1}{2}}\right) \qquad (8.37)$$

where $\hat{F}_{i+\frac{1}{2}}$ is computed using Godunov's method based on the flow variables $\hat{Q}^l_{i+\frac{1}{2}}$ and $\hat{Q}^r_{i+\frac{1}{2}}$, and similarly for $\hat{F}_{i-\frac{1}{2}}$. It can be shown that the method is second-order accurate in time (Hirsch, 1988).

8.4.2 Accuracy, Consistency, and Convergence

We consider the problem described in Section 2.8. The initial condition is defined by (2.92) with $\epsilon = 0.1$ and the domain is $0 < \kappa x < 2\pi$. The norm, defined by (3.102), is evaluated at $\kappa a_o t = 7$, which is prior to the shock formation. The timestep Δt is determined by

$$\Delta t = \mathcal{C}\Delta t_{CFL} \qquad (8.38)$$

where the Courant number $\mathcal{C} = 0.46$ and Δt_{CFL} is calculated according to (5.8) using the initial condition. The timestep Δt is held in fixed ratio to the grid spacing Δx, and therefore Δt and Δx are decreased by the same ratio. The exact solution $Q_i^{n,e}$ is obtained from (2.85) to (2.88).

The convergence is displayed in Fig. 8.33. The solution converges quadratically[†] to the exact solution for sufficiently small Δt. The computed result (using 100 cells) and exact solution are displayed in Fig. 8.34 for $\kappa a_o t = 7$ and $\mathcal{C} = 0.45$.

8.4.3 Total Variation

We examine the Total Variation of the solution of the problem described in Section 2.8. The initial condition is defined by (2.92) with $\epsilon = 0.1$, and the domain $0 < \kappa x < 2\pi$ is discretized into 100 uniform cells. Periodic boundary conditions are imposed on the left and right boundaries. The timestep is defined by (5.8) with $\mathcal{C} = 0.46$. The Total Variation for ρ, ρu, and ρe are

---

† The line in Fig. 8.33 is

$$||Q_i^{n,e} - Q_i^n||_{\kappa a_o \Delta t} = \left(\frac{\kappa a_o \Delta t}{5\times 10^{-4}}\right)^2 ||Q_i^{n,e} - Q_i^n||_{\kappa a_o \Delta t = 5\times 10^{-4}}$$

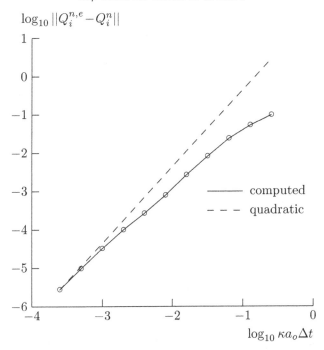

Fig. 8.33. Convergence for MUSCL-Hancock

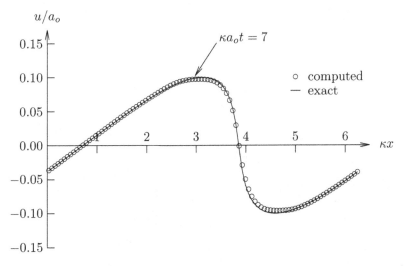

Fig. 8.34. Computed and exact solutions ($\kappa a_o \Delta t = 0.025$ and 100 cells)

shown in Fig. 8.35 compared with the exact solution. The agreement is very good.

The velocity profile is displayed at $\kappa a_o t = 14$ in Fig. 8.36 for $\kappa a_o \Delta t =$

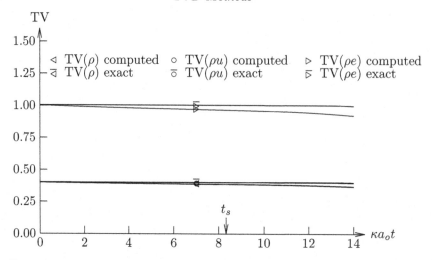

Fig. 8.35. Total Variation of ρ, ρu, and ρe ($\kappa a_o \Delta t = 0.025$ and 100 cells)

0.025 and 100 cells. A decrease in the grid spacing and timestep by a factor of two improves the accuracy of the solution in the vicinity of the shock (Fig. 8.37).

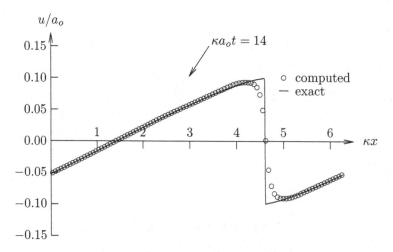

Fig. 8.36. Velocity at $\kappa a_o t = 14$ for MUSCL-Hancock ($\kappa a_o \Delta t = 0.025$ and 100 cells)

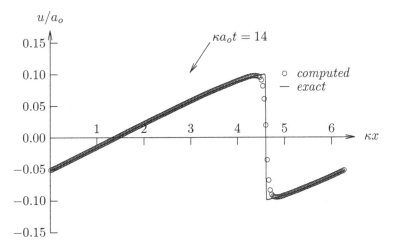

Fig. 8.37. Velocity at $\kappa a_o t = 14$ for MUSCL-Hancock ($\kappa a_o \Delta t = 0.0125$ and 200 cells)

Exercises

8.1 For a periodic function u, show

$$TV(u) = 2 \sum \left(u_{\max_i} - u_{\min_i} \right)$$

where $u_{\max_i}$ and $u_{\min_i}$ are the local maxima and minima of u.

SOLUTION

From (8.5),

$$TV(u) = \int_{x_o}^{x_o + \lambda} \left| \frac{\partial u}{\partial x} \right| dx$$

where x_o is an arbitrary point and λ is the wavelength. Consider a single maximum at x_1 and minimum at x_2 where $x_o < x_1 < x_2 < x_o + \lambda$. Then, by assumption,

$$\frac{\partial u}{\partial x} = \begin{cases} \geq 0 & \text{for } x_o \leq x \leq x_1 \\ \leq 0 & \text{for } x_1 \leq x \leq x_2 \\ \geq 0 & \text{for } x_2 \leq x \leq x_o + \lambda \end{cases}$$

Thus,

$$\begin{aligned} TV &= \int_{x_o}^{x_1} \frac{\partial u}{\partial x} dx - \int_{x_1}^{x_2} \frac{\partial u}{\partial x} dx + \int_{x_2}^{x_o + \lambda} \frac{\partial u}{\partial x} dx \\ &= u(x_1) - u(x_o) - (u(x_2) - u(x_1)) + u(x_o + \lambda) - u(x_2) \\ &= 2 \left(u(x_1) - u(x_2) \right) \end{aligned}$$

since $u(x_o) = u(x_o + \lambda)$ from periodicity. The above result can also be shown to be true if $x_2 < x_1$. Now assume that there are two local maxima (and therefore two local minima) in $x_o \leq x \leq x_o + \lambda$. The interval may be subdivided into two intervals in which there is only one local maximum and one local minimum. A straightforward evaluation of the above integral yields (8.3).

8.2   Consider the Riemann Shock Tube Problem (Section 2.10). Assume
       that the tube is of finite length with impermeable ends. Is $TV(\rho)$
       constant for all time $t > 0$?

8.3   Consider the scalar equation

$$\frac{\partial u}{\partial t} + \frac{\partial f}{\partial x} = 0$$

where $f(u)$ is the flux. The equation may be rewritten in semi-
discrete form as

$$\frac{du_i}{dt} = -\frac{1}{\Delta x}\left(\hat{f}_{i+\frac{1}{2}} - \hat{f}_{i-\frac{1}{2}}\right)$$

where $\hat{f}_{i+\frac{1}{2}}(u_i, u_{i+1})$ is the discrete flux at $x_{i+\frac{1}{2}}$ and is assumed to
depend only on u_i and u_{i+1}. Therefore, we may write

$$\hat{f}_{i+\frac{1}{2}} - \hat{f}_{i-\frac{1}{2}} = C^{-}_{i+\frac{1}{2}}\Delta u_{i+\frac{1}{2}} + C^{+}_{i-\frac{1}{2}}\Delta u_{i-\frac{1}{2}}$$

where

$$\Delta u_{i+\frac{1}{2}} = u_{i+1} - u_i$$

and $C^{-}_{i+\frac{1}{2}}$ and $C^{+}_{i-\frac{1}{2}}$ are coefficients that may depend on u_{i-1}, u_i and
u_{i+1}. Show that the necessary conditions for the semi-discrete form
to be TVD are (Hirsch, 1988)

$$C^{-}_{i+\frac{1}{2}} \leq 0 \quad \text{and} \quad C^{+}_{i+\frac{1}{2}} \geq 0$$

SOLUTION

Using

$$\frac{du_i}{dt} = -\frac{1}{\Delta x}\left(C^{-}_{i+\frac{1}{2}}\Delta u_{i+\frac{1}{2}} + C^{+}_{i-\frac{1}{2}}\Delta u_{i-\frac{1}{2}}\right)$$

thus

$$\frac{d}{dt}(u_{i+1} - u_i) = -\frac{1}{\Delta x}\left(C^{-}_{i+\frac{3}{2}}\Delta u_{i+\frac{3}{2}} + C^{+}_{i+\frac{1}{2}}\Delta u_{i+\frac{1}{2}} - C^{-}_{i+\frac{1}{2}}\Delta u_{i+\frac{1}{2}} - C^{+}_{i-\frac{1}{2}}\Delta u_{i-\frac{1}{2}}\right)$$

The discrete expression for $TV(u)$ is

$$TV(u) = \sum_i |u_{i+1} - u_i|$$

and thus

$$\frac{dTV(u)}{dt} = \sum_i s_{i+\frac{1}{2}}\frac{d}{dt}(u_{i+1} - u_i)$$

where $s_{i+\frac{1}{2}} = \text{sign}(u_{i+1} - u_i)$. For any function g,

$$\frac{d}{dt}|g| = \text{sign}\, g\,\frac{dg}{dt}$$

except at $g = 0$, where $d|g|/dt$ is possibly undefined. Thus,

$$\frac{d\,TV(u)}{dt} = \frac{1}{\Delta x} \sum_i \left\{ \left(C^-_{i+\frac{1}{2}} - C^+_{i+\frac{1}{2}} \right) \Delta u_{i+\frac{1}{2}} - C^-_{i+\frac{3}{2}} \Delta u_{i+\frac{3}{2}} + C^+_{i-\frac{1}{2}} \Delta u_{i-\frac{1}{2}} \right\} s_{i+\frac{1}{2}}$$

Assume a periodic domain. Then

$$\sum_i C^-_{i+\frac{3}{2}} \Delta u_{i+\frac{3}{2}} s_{i+\frac{1}{2}} = \sum_i C^-_{i+\frac{1}{2}} \Delta u_{i+\frac{1}{2}} s_{i-\frac{1}{2}}$$

and

$$\sum_i C^+_{i-\frac{1}{2}} \Delta u_{i-\frac{1}{2}} s_{i+\frac{1}{2}} = \sum_i C^+_{i+\frac{1}{2}} \Delta u_{i+\frac{1}{2}} s_{i+\frac{3}{2}}$$

Thus,

$$\frac{d\,TV(u)}{dt} = \frac{1}{\Delta x} \sum_i \left\{ s_{i+\frac{1}{2}} \left(C^-_{i+\frac{1}{2}} - C^+_{i+\frac{1}{2}} \right) - C^-_{i+\frac{1}{2}} s_{i-\frac{1}{2}} + C^+_{i+\frac{1}{2}} s_{i+\frac{3}{2}} \right\} \Delta u_{i+\frac{1}{2}}$$

Therefore, a sufficient condition for the semi-discrete form to be TVD is either

$$\operatorname{sign} \phi_{i+\frac{1}{2}} = -s_{i+\frac{1}{2}}$$

where

$$\phi_{i+\frac{1}{2}} = s_{i+\frac{1}{2}} \left(C^-_{i+\frac{1}{2}} - C^+_{i+\frac{1}{2}} \right) - C^-_{i+\frac{1}{2}} s_{i-\frac{1}{2}} + C^+_{i+\frac{1}{2}} s_{i+\frac{3}{2}}$$

for all $\Delta u_{i+\frac{1}{2}}$ or

$$\phi_{i+\frac{1}{2}} = 0$$

A total of eight different cases can be considered as indicated in the following table:

Case	$s_{i+\frac{3}{2}}$	$s_{i+\frac{1}{2}}$	$s_{i-\frac{1}{2}}$
1	+1	+1	+1
2	+1	+1	−1
3	+1	−1	+1
4	+1	−1	−1
5	−1	+1	+1
6	−1	+1	−1
7	−1	−1	+1
8	−1	−1	−1

For Case 1, $\phi_{i+\frac{1}{2}} = 0$. For Case 2, $C^-_{i+\frac{1}{2}} \leq 0$. For Case 3, $C^+_{i+\frac{1}{2}} - C^-_{i+\frac{1}{2}} \geq 0$. For Case 4, $C^+_{i+\frac{1}{2}} \geq 0$. Consideration of these and the remaining cases (see Exercise 8.4) yields the necessary conditions

$$C^+_{i+\frac{1}{2}} \geq 0$$
$$C^-_{i+\frac{1}{2}} \leq 0$$

8.4 Evaluate Cases 5 through 8 in Exercise 8.3 and show that the necessary conditions are

$$C^+_{i+\frac{1}{2}} \geq 0$$
$$C^-_{i+\frac{1}{2}} \leq 0$$

8.5   Consider the scalar equation of Exercise 3. Show that

$$C^-_{i+\frac{1}{2}} + C^+_{i+\frac{1}{2}} = [f(u_{i+1}) - f(u_i)](u_i - u_{i-1})^{-1}$$

SOLUTION

Consistency of the discrete flux expression $\hat{f}_{i+\frac{1}{2}}(u_i, u_{i+1})$ requires $\hat{f}(u, u) = f(u)$, where $f(u)$ is the flux. Assume $u_{i-1} = u_i$. Then, the flux difference expression

$$\hat{f}_{i+\frac{1}{2}} - \hat{f}_{i-\frac{1}{2}} = C^-_{i+\frac{1}{2}} \Delta u_{i+\frac{1}{2}} + C^+_{i-\frac{1}{2}} \Delta u_{i-\frac{1}{2}}$$

yields

$$\hat{f}_{i+\frac{1}{2}} - f(u_i) = C^-_{i+\frac{1}{2}} \Delta u_{i+\frac{1}{2}}$$

Similarly, assume $u_i = u_{i+1}$ and obtain

$$f(u_i) - \hat{f}_{i-\frac{1}{2}} = C^+_{i-\frac{1}{2}} \Delta u_{i-\frac{1}{2}}$$

Combining these equations yields

$$C^-_{i+\frac{1}{2}} = \left[\hat{f}_{i+\frac{1}{2}} - f(u_i)\right](u_{i+1} - u_i)^{-1}$$

$$C^+_{i-\frac{1}{2}} = \left[f(u_i) - \hat{f}_{i-\frac{1}{2}}\right](u_i - u_{i-1})^{-1}$$

Combining again,

$$C^-_{i+\frac{1}{2}} + C^+_{i+\frac{1}{2}} = [f(u_{i+1}) - f(u_i)](u_i - u_{i-1})^{-1}$$

8.6   Assume $f = au$ for the scalar equation of Exercise 8.3, where a is a constant. Assume the discrete flux is given by

$$\hat{f}_{i+\frac{1}{2}} = \frac{a}{2}(u_{i+1} + u_i)$$

Is this method TVD?

8.7   Prove that the corrected flux $F^c_{i+\frac{1}{2}}$ has the same sign as the anti-diffusive flux $F^a_{i+\frac{1}{2}}$ [Equation (8.18)].

SOLUTION

From Equation (8.13), if $K_{i+\frac{1}{2}} > 0$, then $F^c_{i+\frac{1}{2}} = S_{i+\frac{1}{2}} K_{i+\frac{1}{2}}$ and hence Equation (8.18) holds. If $K_{i+\frac{1}{2}} \leq 0$, then $F^c_{i+\frac{1}{2}} = 0$.

8.8   Prove $F^c_{i+\frac{1}{2}} \neq 0$ if and only if

$$\text{sign}\left(\Delta \hat{Q}_i\right) = \text{sign}\left(\Delta \hat{Q}_{i+2}\right) = \text{sign}\left(F^a_{i+\frac{1}{2}}\right)$$

8.9   Prove that

$$F^c_{i+\frac{1}{2}} = 0 \quad \text{if} \quad \text{sign}(\Delta\hat{Q}_{i+2}) = -\text{sign}(\Delta\hat{Q}_i)$$

SOLUTION

From (8.14) it is evident that $K_{i+\frac{1}{2}} < 0$ if $\text{sign}(\Delta\hat{Q}_{i+2}) = -\text{sign}(\Delta\hat{Q}_i)$. Therefore, from (8.13)

$$F^c_{i+\frac{1}{2}} = 0 \quad \text{if} \quad \text{sign}(\Delta\hat{Q}_{i+2}) = -\text{sign}(\Delta\hat{Q}_i)$$

8.10   For Class 16 in Flux Corrected Transport, there are four possible cases depending on the sign of $F^a_{i\pm\frac{1}{2}}$. Prove that no new extrema are created relative to the set $(\hat{Q}_{i-2}, \ldots, \hat{Q}_{i+2})$.

Notes

Chapter 2

1 Consider a locally planar shock at rest in the frame of reference (*i.e.*, $u_w = 0$) as shown in Fig. 1. The flow states on the left and right sides are denoted by the subscripts 1 and 2, respectively. The component of the velocity normal to the shock is denoted by u.

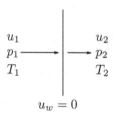

$$u_w = 0$$

Fig. 1   Normal shock at rest

The Rankine-Hugoniot conditions (Table 2.1) yield

$$\rho_1 u_1 = \rho_2 u_2 \tag{8.1}$$

$$\rho_1 u_1^2 + p_1 = \rho_2 u_2^2 + p_2 \tag{8.2}$$

$$h_1 + \tfrac{1}{2} u_1^2 = h_2 + \tfrac{1}{2} u_2^2 \tag{8.3}$$

where in (8.3) the conservation of mass (8.1) and definition of static enthalpy $h = c_p T$ have been used. Equation (8.3) indicates that the total enthalpy $H = h + \frac{1}{2} u^2$ is conserved across the shock, and thus the total temperature $T_o = H/c_p$ is also conserved.

Define the *sonic static temperature* T^* as the static temperature achieved when the flow is compressed or expanded adiabatically to sonic conditions (*i.e.*, at constant total enthalpy). Thus, the sonic temperature T^* and total temperature T_o are related by

$$T_* = \frac{2}{(\gamma + 1)} T_o$$

The sonic speed of sound a_* is defined as

$$a_* = \sqrt{\gamma R T_*}$$

and is conserved across the shock.

From (8.2),

$$\rho_1 u_1^2 + p_1 = \rho_2 u_2^2 + p_2$$

Divide the left side by $\rho_1 u_1$ and the right side by $\rho_2 u_2$ (noting $\rho_1 u_1 = \rho_2 u_2$),

$$\frac{p_1}{\rho_1 u_1} + u_1 = \frac{p_2}{\rho_2 u_2} + u_2$$

Since $a = \sqrt{\gamma R T} = \sqrt{\gamma p / \rho}$,

$$\frac{a_1^2}{\gamma u_1} + u_1 = \frac{a_2^2}{\gamma u_2} + u_2$$

Equation (8.3) may be rewritten as

$$\frac{a_1^2}{(\gamma - 1)} + \tfrac{1}{2} u_1^2 = \frac{(\gamma + 1)}{2(\gamma - 1)} a_*^2$$

and since a_* is conserved across the shock, then

$$a_1^2 = \frac{(\gamma + 1)}{2} a_*^2 - \frac{(\gamma - 1)}{2} u_1^2$$

$$a_2^2 = \frac{(\gamma + 1)}{2} a_*^2 - \frac{(\gamma - 1)}{2} u_2^2$$

and therefore

$$\frac{a_*^2}{u_1} + u_1 = \frac{a_*^2}{u_2} + u_2$$

which yields the *Prandtl relation*

$$u_1 u_2 = a_*^2$$

or, equivalently,

$$M_{*_1} M_{*_2} = 1 \tag{8.4}$$

where $M_* = u / a_*$. Using the definition of a_*,

$$
\begin{aligned}
M_* &= \frac{u}{a} \frac{a}{a_*} \\
&= \frac{u}{a} \frac{a}{a_o} \frac{a_o}{a_*} \\
&= M \left[\frac{T}{T_o} \frac{T_o}{T_*} \right]^{1/2} \\
&= M \left[\left(1 + \frac{(\gamma - 1)}{2} M^2 \right)^{-1} \frac{(\gamma + 1)}{2} \right]^{1/2} \tag{8.5}
\end{aligned}
$$

Thus, from (8.4),

$$M_2^2 = \left[1 + \frac{(\gamma - 1)}{2} M_1^2 \right] \left[\gamma M_1^2 - \frac{(\gamma - 1)}{2} \right]^{-1}$$

From (8.1),

$$\frac{\rho_2}{\rho_1} = \frac{u_1}{u_2}$$

$$= \frac{u_1^2}{u_1 u_2}$$

$$= \frac{u_1^2}{a_*^2}$$

$$= M_{*1}^2 \tag{8.6}$$

using (8.4.3). Thus, from (8.5),

$$\frac{\rho_2}{\rho_1} = \frac{(\gamma+1)}{2} M_1^2 \left[1 + \frac{(\gamma-1)}{2} M_1^2 \right]^{-1}$$

From (8.2),

$$p_2 - p_1 = \rho_1 u_1^2 - \rho_2 u_2^2$$

$$= \rho_1 u_1^2 \left(1 - \frac{u_2}{u_1} \right)$$

using (8.1), and thus

$$\frac{p_2}{p_1} = 1 + \gamma M_1^2 \left(1 - \frac{\rho_1}{\rho_2} \right)$$

and thus

$$\frac{p_2}{p_1} = 1 + \frac{2\gamma}{(\gamma+1)} \left(M_1^2 - 1 \right)$$

From the ideal gas equation,

$$\frac{T_2}{T_1} = \left[1 + \frac{2\gamma}{(\gamma+1)} \left(M_1^2 - 1 \right) \right] \left[\frac{(\gamma-1)M_1^2 + 2}{(\gamma+1)M_1^2} \right]$$

From the definition of entropy,

$$\frac{s_2 - s_1}{c_v} = \ln \left\{ \left[1 + \frac{2\gamma}{(\gamma+1)} \left(M_1^2 - 1 \right) \right] \left[\frac{2 + (\gamma-1)M_1^2}{(\gamma+1)M_1^2} \right]^\gamma \right\}$$

The results for the stationary normal shock may be summarized as follows:

$$\frac{\rho_2}{\rho_1} = \frac{(\gamma+1)M_1^2}{(\gamma-1)M_1^2 + 2}$$

$$\frac{u_1}{u_2} = \frac{\rho_2}{\rho_1}$$

$$\frac{p_2}{p_1} = 1 + \frac{2\gamma}{(\gamma+1)} \left(M_1^2 - 1 \right)$$

$$\frac{T_2}{T_1} = \left[1 + \frac{2\gamma}{(\gamma+1)} \left(M_1^2 - 1 \right) \right] \left[\frac{(\gamma-1)M_1^2 + 2}{(\gamma+1)M_1^2} \right]$$

$$M_2^2 = \frac{(\gamma-1)M_1^2 + 2}{2\gamma M_1^2 - (\gamma-1)}$$

$$\frac{s_2 - s_1}{c_v} = \ln \left\{ \left[1 + \frac{2\gamma}{(\gamma+1)} \left(M_1^2 - 1 \right) \right] \left[\frac{2 + (\gamma-1)M_1^2}{(\gamma+1)M_1^2} \right]^\gamma \right\} \tag{8.7}$$

Additionally, the density and pressure ratio across the shock are related by

$$\frac{\rho_2}{\rho_1} = \frac{(\gamma - 1)p_1 + (\gamma + 1)p_2}{(\gamma + 1)p_1 + (\gamma - 1)p_2} \tag{8.8}$$

Now consider a frame of reference in which the shock is moving with velocity u_w and the normal components of the velocity are u_l and u_r on the left and right sides of the shock, respectively. There are two possible cases as indicated in Fig. 2, where the velocity components are shown in the frame of reference attached to the *shock*.

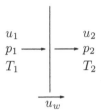

Fig. 2 Normal shock moving with velocity u_w

Case 1: $u_1 - u_w > 0$

The fluid moves from left to right across the shock in the frame of reference of the shock. From the normal shock relations applied in the frame of reference of the shock wave,

$$\frac{p_2}{p_1} = 1 + \frac{2\gamma}{(\gamma + 1)} \left[\left(\frac{u_1 - u_w}{a_1} \right)^2 - 1 \right]$$

and thus

$$u_w = u_1 - a_1 \sqrt{\frac{\gamma + 1}{2\gamma} (\sigma_l - 1) + 1}$$

where $\sigma_l = p_2/p_1$ is the static pressure ratio across the shock. The negative sign is taken for the square root since $u_1 - u_w > 0$. The density and velocity are related by

$$\frac{\rho_2}{\rho_1} = \frac{u_1 - u_w}{u_2 - u_w} \tag{8.9}$$

and furthermore, using (8.8),

$$\frac{\rho_2}{\rho_1} = \frac{(\gamma - 1) + (\gamma + 1)\sigma_l}{(\gamma + 1) + (\gamma - 1)\sigma_l}$$

From the ideal gas equation,

$$\frac{T_2}{T_1} = \frac{\rho_1}{\rho_2} \sigma_l$$

and from (8.9),

$$u_2 = \frac{\rho_2}{\rho_1} \left[u_1 + u_w \left(\frac{\rho_2}{\rho_1} - 1 \right) \right]$$

which can be rewritten as

$$u_2 = u_1 - \frac{a_1}{\gamma} \frac{(\sigma_l - 1)}{\sqrt{\frac{(\gamma+1)}{2\gamma}\sigma_l + \frac{(\gamma-1)}{2\gamma}}} \tag{8.10}$$

Case 2: $u_1 - u_w < 0$

The fluid moves from right to left across the shock in the frame of reference of the shock. From the normal shock relations,

$$\frac{p_1}{p_2} = 1 + \frac{2\gamma}{(\gamma + 1)}\left[\left(\frac{u_w - u_2}{a_2}\right)^2 - 1\right]$$

and thus

$$u_w = u_2 + a_2\sqrt{\frac{\gamma + 1}{2\gamma}(\sigma_r - 1) + 1}$$

where $\sigma_r = p_1/p_2$ is the static pressure ratio across the shock. The positive sign is taken for the square root since $u_1 - u_w < 0$. The density and velocity are related by

$$\frac{\rho_1}{\rho_2} = \frac{u_w - u_2}{u_w - u_1} \tag{8.11}$$

and furthermore, using (8.8),

$$\frac{\rho_1}{\rho_2} = \frac{(\gamma - 1) + (\gamma + 1)\sigma_r}{(\gamma + 1) + (\gamma - 1)\sigma_r}$$

From the ideal gas equation,

$$\frac{T_1}{T_2} = \frac{\rho_2}{\rho_1}\sigma_r$$

and from (8.11),

$$u_1 = \frac{\rho_2}{\rho_1}\left[u_2 + u_w\left(\frac{\rho_1}{\rho_2} - 1\right)\right]$$

which can be rewritten as

$$u_1 = u_2 + \frac{a_2}{\gamma}\frac{(\sigma_r - 1)}{\sqrt{\frac{(\gamma+1)}{2\gamma}\sigma_r + \frac{(\gamma-1)}{2\gamma}}} \tag{8.12}$$

2 The solution consists of four regions divided by the left shock, the contact surface, and the right shock. The compatibility conditions are

$$u_2 = u_3$$
$$p_2 = p_3$$

Using (2.46) and (2.51), the first condition becomes

$$u_1 - a_1 f(p^*, p_1) = u_4 + a_4 f(p^*, p_4)$$

where p^* denotes the static pressure at the contact surface (and hence $p_2 = p_3 = p^*$), and

$$f(p^*, p) = \frac{1}{\gamma}\left(\frac{p}{p^*} - 1\right)\left[\frac{(\gamma + 1)}{2\gamma}\frac{p^*}{p} + \frac{(\gamma - 1)}{2\gamma}\right]^{-1/2}$$

which correspond to (2.98) and (2.99) for $p^* > p_1$ and $p^* > p_4$. Equations (2.100) to (2.108) follow from the results in Section 2.3.

3 The solution consists of four regions divided by the left shock, the contact surface, and the right expansion. The compatibility conditions are

$$u_2 = u_3$$
$$p_2 = p_3$$

Using (2.46) and (2.72), the first condition becomes

$$u_1 - a_1 f(p^*, p_1) = u_4 + a_4 f(p^*, p_4)$$

where p^* denotes the static pressure at the contact surface (and hence $p_2 = p_3 = p^*$). Since $p^* > p_1$, the expression for $f(p^*, p)$ on the left side of the equation is

$$f(p^*, p) = \frac{1}{\gamma} \left(\frac{p}{p^*} - 1 \right) \left[\frac{(\gamma + 1)}{2\gamma} \frac{p^*}{p} + \frac{(\gamma - 1)}{2\gamma} \right]^{-1/2}$$

Since $p^* < p_4$, the expression for $f(p^*, p)$ on the right side of the equation is

$$f(p^*, p) = \frac{2}{(\gamma - 1)} \left[\left(\frac{p^*}{p} \right)^{(\gamma - 1)/2\gamma} - 1 \right]$$

Equations (2.109) to (2.121) follow from the results in Section 2.3.

4 The solution consists of four regions divided by the left expansion, the contact surface, and the right shock. The compatibility conditions are

$$u_2 = u_3$$
$$p_2 = p_3$$

Using (2.51) and (2.68), the first condition becomes

$$u_1 - a_1 f(p^*, p_1) = u_4 + a_4 f(p^*, p_4)$$

where p^* denotes the static pressure at the contact surface (and hence $p_2 = p_3 = p^*$). Since $p^* < p_1$, the expression for $f(p^*, p)$ on the left side of the equation is

$$f(p^*, p) = \frac{2}{(\gamma - 1)} \left[\left(\frac{p^*}{p} \right)^{(\gamma - 1)/2\gamma} - 1 \right]$$

Since $p^* > p_4$, the expression for $f(p^*, p)$ on the right side of the equation is

$$f(p^*, p) = \frac{1}{\gamma} \left(\frac{p}{p^*} - 1 \right) \left[\frac{(\gamma + 1)}{2\gamma} \frac{p^*}{p} + \frac{(\gamma - 1)}{2\gamma} \right]^{-1/2}$$

Equations (2.122) to (2.134) follow from the results in Section 2.3.

5 The solution consists of four regions divided by the left expansion, the contact surface, and the right expansion. The compatibility conditions are

$$u_2 = u_3$$
$$p_2 = p_3$$

Using (2.68) and (2.72), the first condition becomes

$$u_1 - a_1 f(p^*, p_1) = u_4 + a_4 f(p^*, p_4)$$

where p^* denotes the static pressure at the contact surface (and hence $p_2 = p_3 = p^*$).

Since $p^* < p_1$ and $p^* < p_4$, the expression for $f(p^*, p)$ on the both sides of the equation is

$$f(p^*, p) = \frac{2}{(\gamma - 1)} \left[\left(\frac{p^*}{p} \right)^{(\gamma-1)/2\gamma} - 1 \right]$$

Equations (2.135) to (2.151) follow from the results in Section 2.3.

Chapter 3

1 Inconsistent discrete approximations are relatively rare in computational science, although they are certainly possible to create. The classic example is the DuFort-Frankel (1953) algorithm for the heat equation

$$\frac{\partial T}{\partial t} = \kappa \frac{\partial^2 T}{\partial x^2} \tag{8.13}$$

Assuming the temperature T is defined at a discrete set of equally spaced points x_i, the algorithm is

$$\frac{(T_i^{n+1} - T_i^{n-1})}{\Delta t} = \kappa \frac{(T_{i+1}^n - T_i^{n+1} - T_i^{n-1} + T_{i-1}^{n-1})}{\Delta x^2} \tag{8.14}$$

Expanding each term in a Taylor series about x_i at t^n yields

$$\frac{(T_i^{n+1} - T_i^{n-1})}{\Delta t} - \kappa \frac{(T_{i+1}^n - T_i^{n+1} - T_i^{n-1} + T_{i-1}^{n-1})}{\Delta x^2} =$$

$$\frac{\partial T}{\partial t} - \kappa \frac{\partial^2 T}{\partial x^2} + \mathcal{E}(\Delta x, \Delta t)$$

where

$$\mathcal{E}(\Delta x, \Delta t) = \kappa \left(\frac{\Delta t}{\Delta x} \right)^2 \frac{\partial^2 T}{\partial t^2} + \mathcal{O}(\Delta t)^2 + \mathcal{O}(\Delta x^2) + \mathcal{O} \left(\frac{\Delta t^4}{\Delta x^2} \right) \tag{8.15}$$

The discrete approxiation (8.14) is consistent with (8.13) provided

$$\lim_{\substack{\Delta x \to 0 \\ \Delta t \to 0}} \frac{\Delta t}{\Delta x} = 0$$

This can be achieved, for example, if $\Delta t \propto \kappa^{-1} \Delta x^2$. For a grid refinement study employing a succession of smaller spatial grid spacings,

$$\Delta x_1 > \Delta x_2 > \ldots > \Delta x_j > \ldots$$

(*e.g.*, $\Delta x_1 = \Delta x$, $\Delta x_2 = \frac{1}{2}\Delta x$, ...), and a corresponding sequence of smaller timesteps

$$\Delta t_1 > \Delta t_2 > \ldots > \Delta t_j > \ldots$$

this implies that the corresponding timestep must decrease according to

$$\frac{\Delta t_{j+1}}{\Delta t_j} = \left(\frac{\Delta x_{j+1}}{\Delta x_j} \right)^2 \quad \text{for} \quad j = 1, 2, \ldots$$

Conversely, if the timestep was selected according to $\Delta t = c^{-1}\Delta x$, where c is a constant having dimensions of velocity, then for $\Delta x \to 0$ and $\Delta t \to 0$ the discrete approximation would be consistent with

$$\frac{\partial^2 T}{\partial t^2} - c^2 \frac{\partial^2 T}{\partial x^2} + \frac{c^2}{\kappa}\frac{\partial T}{\partial t} = 0 \tag{8.16}$$

which is a hyperpolic equation with fundamentally different properties than (8.13).

2 The Von Neumann Method for stability analysis was developed by John von Neumann at Los Alamos during World War II and was originally classified. It was later declassified and published in Crank and Nicholson (1947) and Charney *et al.* (1950) (Hirsch, 1988).

3 The norm of a vector x is a measure of the length of the vector. The norm is denoted $|x|$ and satisfies the following properties (Isaacson and Keller, 1966):

(i) For each vector x, there is a unique real norm $|x|$.
(ii) The norm is nonnegative for all x and $|x| = 0$ if and only if $x = 0$.
(iii) For any scalar α and any x,

$$|\alpha x| = |\alpha||x|$$

(iv) For any vectors x and y,

$$|x + y| \le |x| + |y|$$

which is the triangle inequality.

For a vector x with elements $x_i, i = 1, \ldots, n$, examples of norms are

$$|x|_1 = \sum_{i=1}^{i=n} |x_i|$$

$$|x|_2 = \left(\sum_{i=1}^{i=n} |x_i|^2\right)^{\frac{1}{2}}$$

$$|x|_p = \left(\sum_{i=1}^{i=n} |x_i|^p\right)^{\frac{1}{p}}$$

$$|x|_\infty = \max_i |x_i|$$

where $|x|_2$ is recognized as the Euclidean norm. The proof of the triangle inequality for these norms is presented in Isaacson and Keller (1966).

4 The norm of a matrix A is a measure of the "stretching" achieved by multiplication of a vector x by the matrix. Thus,

$$\|A\| = \sup_{x\neq 0} \frac{\|Ax\||}{\|x\|}$$

where $\|\cdot\|$ is a suitable vector norm (see above). This definition implies (Isaacson and Keller, 1966)

$$\|Ax\| \le \|A\|\,\|x\|$$
$$\|A + B\| \le \|A\| + \|B\|$$
$$\|AB\| \le \|A\|\,\|B\|$$

5 A classic method is the two-step algorithm of Lax-Wendroff (Lax and Wendroff, 1960; Richtmyer and Morton, 1967). Originally developed as a finite-difference algorithm, it can be readily interpreted as a finite-volume method using two sets of control volumes. The first set of control volumes is a uniform discretization of the x-axis at time t^n into M cells of length Δx with centroids x_i, $i = 1, \ldots, M$ as illustrated in Fig. 3. The volume-averaged vector of dependent variables on the first set of control volumes is defined by

$$Q_i(t) = \frac{1}{V_i} \int_{V_i} Q \, dx \, dy$$

The second set of control volumes is a uniform discretization of the x-axis at time $t^{n+\frac{1}{2}} = t^n + \frac{1}{2}\Delta t$ into M cells of the same length Δx with centroids at $x_{i+\frac{1}{2}}$, *i.e.*, the cells are shifted an amount $\frac{1}{2}\Delta x$ with respect to the first set. The volume-averaged vector of dependent variables on the first set of control volumes is defined by

$$Q_{i'}(t) = \frac{1}{V_{i'}} \int_{V_{i'}} Q \, dx \, dy$$

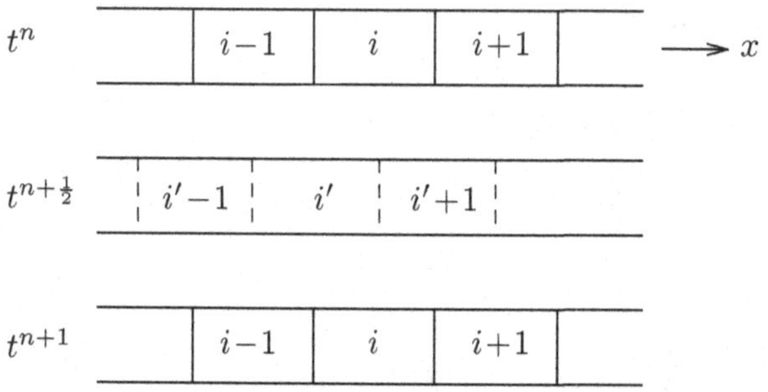

Fig. 3   Alternating grids for Lax-Wendroff

The first step of Lax-Wendroff is equivalent (to second order in Δx) to an explicit Euler integration of (3.1) on the second set of control volumes V_i' for a time interval $\frac{1}{2}\Delta t$:

$$Q_{i'}^{n+\frac{1}{2}} = Q_{i'}^n - \frac{\Delta t}{2\Delta x} \left(F_{i'+\frac{1}{2}}^n - F_{i'-\frac{1}{2}}^n \right)$$

where, to $\mathcal{O}(\Delta x)^2$,

$$Q_{i'}^n = \tfrac{1}{2} \left(Q_i^n + Q_{i-1}^n \right)$$

and the flux is evaluated by

$$F_{i'+\frac{1}{2}}^n = F(Q_i^n)$$

The second step of Lax-Wendroff is equivalent to an explicit Euler integration of (3.1) on the first set of control volumes V_i for a time interval Δt:

$$Q_i^{n+1} = Q_i^n - \frac{\Delta t}{\Delta x} \left(F_{i+\frac{1}{2}}^{n+\frac{1}{2}} - F_{i-\frac{1}{2}}^{n+\frac{1}{2}} \right)$$

where the flux is evaluated using the results of the first step at $t^{n+\frac{1}{2}}$:

$$F_{i+\frac{1}{2}}^{n+\frac{1}{2}} = F(Q_{i'+1}^{n+\frac{1}{2}})$$

Thus, the algorithm is

$$Q_{i'}^{n+\frac{1}{2}} = \frac{1}{2}(Q_i^n + Q_{i-1}^n) - \frac{\Delta t}{2\Delta x}(F(Q_i^n) - F(Q_{i-1}^n)) \qquad (8.17)$$

$$Q_i^{n+1} = Q_i^n - \frac{\Delta t}{\Delta x}\left(F(Q_{i'+1}^{n+\frac{1}{2}}) - F(Q_{i'}^{n+\frac{1}{2}})\right) \qquad (8.18)$$

The algorithm is second-order accurate in x and t (for fixed ratio $\Delta t/\Delta x$) and consistent.

The stability can be determined using the von Neumann method as described in Section 3.8.1. Expanding (8.17) in a Taylor series and retaining the lowest order terms,

$$Q_{i'}^{n+\frac{1}{2}} = \frac{1}{2}(Q_i^n + Q_{i-1}^n) - \frac{\Delta t}{2\Delta x}A(Q_i^n - Q_{i-1}^n)$$

where A is the Jacobian matrix (2.8) defined in Section 2.2.2 and is assumed constant. Thus,

$$Q_{i'}^{n+\frac{1}{2}} = \frac{1}{2}\left[\left(I - \frac{\Delta t}{\Delta x}A\right)Q_i^n + \left(I + \frac{\Delta t}{\Delta x}A\right)Q_{i-1}^n\right]$$

Expanding (8.18) in a similar manner and substituting yields

$$Q_i^{n+1} = -\frac{\Delta t}{2\Delta x}A\left(I - \frac{\Delta t}{\Delta x}A\right)Q_{i+1}^n + \left[I - \left(\frac{\Delta t}{\Delta x}\right)^2 A^2\right]Q_i^n$$
$$+\frac{\Delta t}{2\Delta x}A\left(I + \frac{\Delta t}{\Delta x}A\right)Q_{i-1}^n$$

Substituting the Fourier series

$$Q(x,t) = \sum_{l=-N+1}^{l=N} \hat{Q}_k(t)e^{\iota k x}$$

into the previous equation, where $\iota = \sqrt{-1}$ and $k = 2\pi l L^{-1}$. Thus

$$\hat{Q}_k^{n+1} = G\hat{Q}_k^n$$

where the amplification matrix is

$$G = I + \left(\frac{\Delta t}{\Delta x}\right)^2 (\cos k\Delta x - 1)A^2 - \iota\frac{\Delta t}{\Delta x}\sin k\Delta x A$$

The von Neumann stability condition is

$$|\lambda_{G_i}| \le 1$$

where λ_{G_i} are the eigenvalues of G. Since G is a rational function of A, its eigenvalues are (Protter and Morrey, 1964)

$$\lambda_{G_i} = 1 + \alpha_i^2(\cos k\Delta x - 1) - \iota\alpha_i \sin k\Delta x$$

where

$$\alpha_i = \frac{\Delta t \lambda_i}{\Delta x}$$

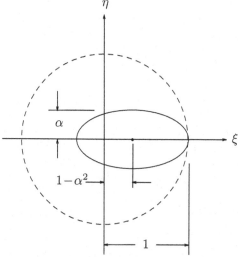

Fig. 4   Stability diagram for Lax-Wendroff

Let

$$\xi = \text{Real}(\lambda_{G_i})$$
$$\eta = \text{Imag}(\lambda_{G_i})$$

Then (8.4.3) is

$$\left[\frac{\xi - (1 - \alpha^2)}{\alpha^2}\right]^2 + \left[\frac{\eta}{\alpha}\right]^2 \leq 1$$

This represents the interior of an ellipse with center at $(\xi, \eta) = (1 - \alpha^2, 0)$, semi-major axis α^2, and semi-minor axis α. The ellipse is shown in Fig. 4. The ellipse lies within the unit circle $|\lambda_{G_i}| = 1$ provided

$$|\alpha_i| \leq 1$$

which is

$$\Delta t \leq \min_i \frac{\Delta x}{|\lambda_i|}$$

The results for the sample problem of Section 3.8.1 are shown in Fig. 5.

Chapter 4

1 The results for $\kappa = -1, 0$, and 1 are shown in Figs. 6 to 8. The reconstructed left and

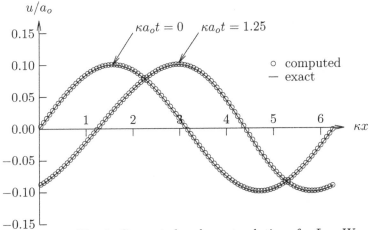

Fig. 5 Computed and exact solutions for Lax-Wendroff

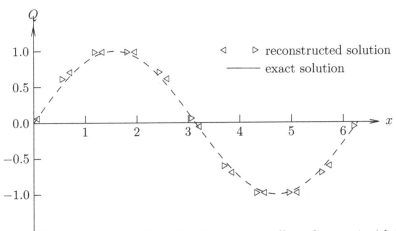

Fig. 6 Reconstruction of sine using ten cells and $\kappa = -1$ with limiter

right states are less accurate in comparison with the exact solution than for $\kappa = \frac{1}{3}$. This is expected due to the lower order of accuracy.

2 The following identity is used in deriving (4.44):

$$Q_i = Q_{i-a} + a\Delta Q_{i-a+\frac{1}{2}} + \frac{a(a-1)}{2}\left(\Delta Q_{i-a+\frac{3}{2}} - \Delta Q_{i-a+\frac{1}{2}}\right)$$

This follows from

$$\frac{1}{\Delta x}\int_{x_{i-\frac{1}{2}}}^{x_{i+\frac{1}{2}}} Q(x)\,dx = Q_i$$

by construction. It may be verified directly for $a = 0, 1$, or 2, which are the only allowable values of a.

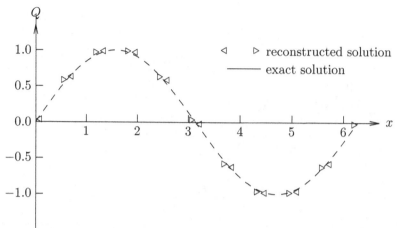

Fig. 7 Reconstruction of sine using ten cells and $\kappa = 0$ with limiter

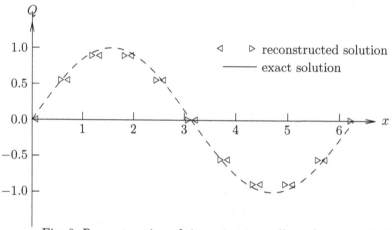

Fig. 8 Reconstruction of sine using ten cells and $\kappa = 1$ with limiter

Chapter 5

1 Godunov's First Method is based on the definition

$$Q_i^{n+1} = \frac{1}{\Delta x} \int_{V_i} \mathcal{Q}(t^{n+1}) dx$$

Consider the piecewise constant solution Q_i for $i = 1, \ldots, M$ at t^n. The left and right states for $\mathcal{Q}$ at $x_{i+\frac{1}{2}}$ may be taken to be Q_i^n and Q_{i+1}^n, respectively. The general Riemann problem is solved for each face as shown in Fig. 9. Provided that there is no intersection of waves from the left and right face Riemann solutions at t^{n+1}, the solution for $\mathcal{Q}$ is therefore known at t^{n+1} as a combination of two general Riemann solutions and (possibly) an intermediate uniform region. The solution for Q_i^{n+1} is therefore

$$Q_i^{n+1} = \frac{1}{\Delta x} \left[\int_{x_{i-\frac{1}{2}}}^{x_l} \mathcal{Q}_{i-\frac{1}{2}}^R \, dx + \int_{x_l}^{x_r} Q_i^n \, dx + \int_{x_r}^{x_{i+\frac{1}{2}}} \mathcal{Q}_{i+\frac{1}{2}}^R \, dx \right] \tag{8.19}$$

where $\mathcal{Q}_{i-\frac{1}{2}}^R$ and $\mathcal{Q}_{i+\frac{1}{2}}^R$ are the solutions to the general Riemann problems at $x_{i-\frac{1}{2}}$ and $x_{i+\frac{1}{2}}$, respectively, and x_l and x_r are the location of the rightmost wave from $\mathcal{Q}_{i-\frac{1}{2}}^R$ and leftmost wave from $\mathcal{Q}_{i+\frac{1}{2}}^R$ at t_{n+1}, respectively. The appropriate modifications when $x_l < x_{i-\frac{1}{2}}$ or $x_r > x_{i+\frac{1}{2}}$ are obvious. The method implies that $x_l < x_r$, *i.e.*, the wave systems of the two general Riemann problems do not intersect. This implies a constraint on the allowable timestep Δt, which may be estimated by

$$\Delta t = \min_i \frac{\Delta x}{2 \, c_{\max}} \tag{8.20}$$

where $c_{\max}$ is the maximum absolute wave speed of the waves entering cell i.

This method is less appealing than Godunov's Second Method because the integration in (8.19) requires more cputime.

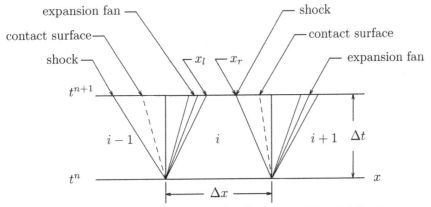

Fig. 9 Possible flow structure for Godunov's First Method

2 The vector R is

$$R_1 = \rho \left[1 - \frac{1}{\gamma} \frac{a^2}{\tilde{a}^2} - \frac{(\gamma - 1)}{2} \frac{(\tilde{u} - u)^2}{\tilde{a}^2} \right]$$

$$R_2 = \rho \left[\frac{(\gamma - 1)}{4} \frac{(u - \tilde{u})^2}{\tilde{a}^2} + \frac{(u - \tilde{u})}{2\tilde{a}} + \frac{1}{2\gamma} \frac{a^2}{\tilde{a}^2} \right]$$

$$R_3 = \rho \left[\frac{(\gamma - 1)}{4} \frac{(u - \tilde{u})^2}{\tilde{a}^2} - \frac{(u - \tilde{u})}{2\tilde{a}} + \frac{1}{2\gamma} \frac{a^2}{\tilde{a}^2} \right]$$

and similarly

$$\rho = R_1 + R_2 + R_3$$
$$\rho u = \tilde{u} R_1 + (\tilde{u} + \tilde{a}) R_2 + (\tilde{u} - \tilde{a}) R_3$$
$$\rho e = \tfrac{1}{2} \tilde{u}^2 R_1 + (\tilde{H} + \tilde{u}\tilde{a}) R_2 + (\tilde{H} - \tilde{u}\tilde{a}) R_3$$

3 An eigenvector r_k of $\mathcal{A}$ is an eigenvector of $\mathcal{A}^+$ corresponding to the eigenvalue λ_k^+ if

$$\mathcal{A}^+ r_k = \lambda_k^+ r_k$$

Writing

$$\mathcal{A}^+ = T\Lambda^+ T^{-1}$$

$$= \underbrace{\left\{ \begin{array}{ccc} r_1 & r_2 & r_3 \end{array} \right\}}_{T} \Lambda^+ \underbrace{\left\{ \begin{array}{c} l_1 \\ l_2 \\ l_3 \end{array} \right\}}_{T^{-1}}$$

where r_k are the column vectors in (2.19) and l_k are the row vectors in (2.20). From (2.21),

$$T^{-1} r_k = e_k$$

where the unit vectors e_k are defined as

$$e_1 = \left\{ \begin{array}{c} 1 \\ 0 \\ 0 \end{array} \right\}, \quad e_2 = \left\{ \begin{array}{c} 0 \\ 1 \\ 0 \end{array} \right\}, \quad e_3 = \left\{ \begin{array}{c} 0 \\ 0 \\ 1 \end{array} \right\}$$

Also,

$$T\Lambda^+ = \left\{ \begin{array}{ccc} \lambda_1^+ r_1 & \lambda_2^+ r_2 & \lambda_3^+ r_3 \end{array} \right\}$$

Therefore,

$$T\Lambda^+ T^{-1} r_k = \left\{ \begin{array}{ccc} \lambda_1^+ r_1 & \lambda_2^+ r_2 & \lambda_3^+ r_3 \end{array} \right\} e_k$$
$$= \lambda_k^+ r_k$$

Similarly, r_k is an eigenvector of $\mathcal{A}^-$ corresponding to the eigenvalue λ_k^-.

4 It is necessary to investigate the condition

$$a_2 \leq c_c$$

at the interfaces $i \pm \frac{1}{2}$. For the left expansion,

$$u_1 + \frac{2}{(\gamma-1)} a_1 = u_2 + \frac{2}{(\gamma-1)} a_2$$

Furthermore,

$$u_2 = c_c$$

Thus,

$$a_2 = a_1 + \frac{(\gamma-1)}{2}(u_1 - c_c)$$

Therefore, the condition

$$a_2 \leq c_c$$

becomes

$$u_1 + \frac{2}{(\gamma-1)} a_1 \leq \frac{(\gamma+1)}{(\gamma-1)} c_c$$

From the solution in Regions 2 and 3,

$$u_1 + \frac{2}{(\gamma-1)}a_1 = c_c + \frac{2}{(\gamma-1)}a_1 \left(\frac{p^*}{p_1}\right)^{(\gamma-1)/2\gamma}$$

Noting that $p^* = p_2$ and

$$\frac{p_2}{p_1} = \left(\frac{a_2}{a_1}\right)^{2\gamma/(\gamma-1)}$$

we obtain

$$c_c \geq a_2$$

But since $c_c = u_2$,

$$u_2 - a_2 \geq 0$$

This is the same condition assumed for the cells i and $i-1$ and therefore may be expected to hold.

Chapter 8

1 The diffusive character of $F^l_{i+\frac{1}{2}}$ can be illustrated for the linear case $F = aQ$ where $a > 0$ is a constant. Using a simple first-order upwind reconstruction,

$$\begin{aligned}
F^l_{i+\frac{1}{2}} &= aQ_i \\
&= aQ_{i+\frac{1}{2}} - a\frac{\Delta x}{2}\left.\frac{\partial Q}{\partial x}\right|_{i+\frac{1}{2}} + \mathcal{O}(\Delta x^2)
\end{aligned} \tag{8.21}$$

Thus, to first order, the semi-discrete equation becomes

$$\frac{dQ_i}{dt} + \frac{\left(aQ_{i+\frac{1}{2}} - aQ_{i-\frac{1}{2}}\right)}{\Delta x} = \Delta x^{-1}\left(\left.\nu\frac{\partial Q}{\partial x}\right|_{i+\frac{1}{2}} - \left.\nu\frac{\partial Q}{\partial x}\right|_{i-\frac{1}{2}}\right) \tag{8.22}$$

where $\nu = a\Delta x/2$. This equation is analogous to

$$\frac{\partial Q}{\partial t} + \frac{\partial aQ}{\partial x} = \nu\frac{\partial^2 Q}{\partial x^2} \tag{8.23}$$

indicating the diffusive effect of a first-order upwind reconstruction for the flux. The low order flux $F^l_{i+\frac{1}{2}}$ therefore generates an anti-diffusive contribution to the corrected flux:

$$F^c_{i+\frac{1}{2}} = \alpha_{i+\frac{1}{2}}\left(F^h_{i+\frac{1}{2}} - F^l_{i+\frac{1}{2}}\right) \tag{8.24}$$

Bibliography

Abramowitz, M. and Stegun, I. (1971) *Handbook of Mathematical Functions*, volume 55. National Bureau of Standards, Washington, DC. Applied Mathematics Series.

Ahlfors, L. (1953) *Complex Analysis*. McGraw-Hill, New York.

Anderson, D., Tannehill, J. and Pletcher, R. (1984) *Computational Fluid Mechanics and Heat Transfer*. Hemisphere Publishing Corporation, New York.

Anderson, W., Thomas, J. and Van Leer, B. (1986) Comparison of Finite Volume Flux Vector Splittings for the Euler Equations. *AIAA Journal*, 24:1453–1460.

Beam, R. and Warming, R. (1976) An Implicit Finite-Difference Algorithm for Hyperbolic Systems in Conservation Law Form. *Journal of Computational Physics*, 22:87–110.

Book, D., Boris, J. and Hain, K. (1975) Flux-Corrected Transport II: Generalizations of the Method. *Journal of Computational Physics*, 18:248–283.

Boris, J. and Book, D. (1973) Flux-Corrected Transport. I. SHASTA, A Fluid Transport Algorithm That Works. *Journal of Computational Physics*, 11:38–69.

Boris, J. and Book, D. (1976) Flux-Corrected Transport III: Minimal-Error FCT Algorithms. *Journal of Computational Physics*, 20:397–431.

Briley, R. and McDonald, H. (1973) Solution of the Three-Dimensional Compressible Navier-Stokes Equations by an Implicit Technique. In *Proceedings of the Fourth International Conference on Numerical Methods in Fluid Dynamics*, pages 105–110, Springer-Verlag, New York.

Chandrasekhara, S. (1981) *Hydrodynamic and Hydromagnetic Stability*. Dover, New York.

Charney, J., Fjortoft, R. and Von Neumann, J. (1950) Numerical Integration of the Barotropic Vorticity Equation. *Tellus*, 2:237–254.

Chu, C.-W. (1997) Essentially Non-Oscillatory and Weighted Essentially Non-Oscillatory Schemes for Hyperbolic Conservation Laws. Technical Report NASA/CR-97-206253 (ICASE Report No. 97-65), NASA Langley Research Center.

Colella, P. and Woodward, P. (1984) The Piecewise Parabolic Method (PPM) for Gas-Dynamical Simulations. *Journal of Computational Physics*, 54:174–201.

Courant, R. and Friedrichs, K. (1948) *Supersonic Flow and Shock Waves*. Springer-Verlag, New York.

Crank, J and Nicholson, P. (1947) A Practical Method for Numerical Evaluation of Solutions of Partial Differential Equations of the Heat Conduction Type. *Proceedings of the Cambridge Philosophical Society*, 43:50–67.

DuFort, E. and Frankel, S. (1953) Stability Conditions in the Numerical Treatment of Parabolic Differential Equations. *Mathematical Tables and Other Aids to Computation*, 7:135–152.

Engquist, B. and Osher, S. (1980) One Sided Difference Approximations for Nonlinear Conservation Laws. *Mathematics of Computation*, 36:321–351.

Fletcher, C. (1988) *Computational Techniques for Fluid Dynamics*, volume I and II. Springer-Verlag, New York.

Franklin, J. (1968) *Matrix Theory.* Prentice-Hall, Inc., Englewood Cliffs, NJ.

Garabedian, P. (1964) *Partial Differential Equations*. John Wiley and Sons, New York.

Gear, C. W. (1971) *Numerical Initial Value Problems in Ordinary Differential Equations*. Prentice-Hall, Englewood Cliffs, NJ.

Godunov, S. (1959) A Finite Difference Method for the Computation of Discontinuous Solutions of the Equations of Fluid Dynamics. Математический Сборник, 47:357–393. In Russian.

Godunov, S. (1999) Reminiscences About Difference Schemes. *Journal of Computational Physics*, 153:6–25.

Gottlieb, J. and Groth, C. (1988) Assessment of Riemann Solvers for Unsteady One-Dimensional Inviscid Flows of Perfect Gases. *Journal of Computational Physics*, 78:437–458.

Greenberg, M. (1998) *Advanced Engineering Mathematics*. Prentice-Hall, Englewood, NJ, second edition.

Harten, A. (1983) High Resolution Schemes for Hyperbolic Conservation Laws. *Journal of Computational Physics*, 49:357–393.

Harten, A. (1984) On a Class of High Resolution Total Variation Stable Finite Difference Schemes. *SIAM Journal of Numerical Analysis*, 21:1–23.

Harten, A., Engquist, B., Osher, S. and Chakravarthy, S. (1987) Uniformly High Order Accurate Essentially Non-oscillatory Schemes, III. *Journal of Computational Physics*, 71:231–303.

Harten, A. and Osher, S. (1987) Uniformly High-Order Accurate Nonoscillatory Schemes, I. *SIAM Journal of Numerical Analysis*, 24:279–309.

Hirsch, C. (1988) *Numerical Computation of Internal and External Flows*, volume I and II. John Wiley and Sons, New York.

Hockney, R. and Jesshope, C. (1988) *Parallel Computers 2*. Adam Hilger, Bristol, UK.

Hoffman, K. (1989) *Computational Fluid Dynamics for Engineers*. Education Systems, Austin, TX.

Holt, M. (1984) *Numerical Methods in Fluid Dynamics*. Springer-Verlag, New York.

Isaacson, E. and Keller, H. (1966) *Analysis of Numerical Methods*. John Wiley and Sons, Inc., New York.

Jameson, A., Schmidt, W. and Turkel, E. (1981) Numerical Solutions of the Euler Equations by Finite Volume Methods Using Runge-Kutta Time-Stepping Schemes. AIAA Paper No. 81-1259. AIAA 14th Fluid and Plasma Dynamics Conference, Palo Alto, CA.

Landau, L. and Lifshitz, E. (1959) *Fluid Mechanics*. Pergammon Press, New York.

Laney, C. (1998) *Computational Gas Dynamics*. Cambridge University Press, Cambridge.

Lax, P. (1959) Weak Solutions of Nonlinear Hyperbolic Equations and Their Numerical Computation. *Communications on Pure and Applied Mathematics,* 7:159–193.

Lax, P. and Richtmyer, R. (1956) Survey of the Stability of Linear Finite Difference Equations. *Communications on Pure and Applied Mathematics,* 9:267–293.

Lax, P. and Wendroff, B. (1960) Systems of Conservation Laws. *Communications on Pure and Applied Mathematics,* 13:217–237.

Liepmann, H. and Roshko, A. (1957) *Elements of Gasdynamics.* John Wiley and Sons, New York.

Liu, X.-D., Osher, S. and Chan, T. (1994) Weighted Essentially Non-Oscillatory Schemes. *Journal of Computational Physics,* 115:200–212.

Mathews, J. and Fink, K. (1998) *Numerical Methods Using Matlab.* Prentice-Hall, Upper Saddle River, NJ.

Matsuno, K. editor. (2003) *Parallel Computational Fluid Dynamics: New Frontiers and Multidisciplinary Applications - Proceedings of the Parallel CFD 2002 Conference,* Kansai Science City, Japan.

Morton, K. and Mayers, D. (1994) *Numerical Solution of Partial Differential Equations.* Cambridge University Press, Cambridge.

Osher, S. and Chakravarthy, S. (1983) Upwind Schemes and Boundary Conditions with Applications to Euler Equations in General Coordinates. *Journal of Computational Physics,* 50:447–481.

Osher, S. and Solomon, F. (1982) Upwind Difference Schemes for Hyperbolic Conservation Laws. *Mathematics of Computation,* 38:339–374.

Press, W., Flannery, B., Teukolsky, S. and Vetterling, W. (1986) *Numerical Recipes.* Cambridge University Press, Cambridge.

Protter, M. and Morrey, C. (1964) *Modern Mathematical Analysis.* Addison-Wesley Publishing Co., Reading, MA.

Quirk, J. (1994) An Alternative to Unstructured Grids for Computing Gas Dynamic Flows Around Arbitrarily Complex Two Dimensional Bodies. *Computers and Fluids,* 23:125–142.

Richtmyer, R. and Morton, K. (1967) *Difference Methods for Initial-Value Problems.* Interscience Publishers, New York, second edition.

Roe, P. (1981) Approximate Riemann Solvers, Parameter Vectors, and Difference Schemes. *Journal of Computational Physics,* 43:357–372.

Roe, P. (1986) Characteristic-Based Schemes for the Euler Equations. *Annual Review of Fluid Mechanics,* 18:337–365.

Runge, C. (1912) *Graphical Methods.* Columbia University Press, New York.

Sanders, R. and Prendergast, K. (1974) . *Astrophysical Journal,* 188:489.

Schreier, S. (1982) *Compressible Flow.* John Wiley and Sons, New York.

Shapiro, A. (1953) *The Dynamics and Thermodynamics of Compressible Fluid Flow.* John Wiley and Sons, New York.

Steger, J. and Warming, R. (1981) Flux Vector Splitting of the Inviscid Gasdynamic Equations with Applications to Finite-Difference Methods. *Journal of Computational Physics,* 40:263–293.

Stoer, J. and Bulirsch, R. (1970) *Introduction to Numerical Analysis.* Springer-Verlag, New York.

Toro, E. (1997) *Riemann Solvers and Numerical Methods for Fluid Dynamics.* Springer-Verlag, New York.

Van Leer, B. (1977a) Towards the Ultimate Conservative Difference Scheme III. Upstream-Centered Finite Difference Schemes for Ideal Compressible Flow. *Journal of Computational Physics,* 23:263–275, 1977a.

Van Leer, B. (1977b) Towards the Ultimate Conservative Difference Scheme IV. A New Approach to Numerical Convection. *Journal of Computational Physics*, 23:276–299.

Van Leer, B. (1979) Towards the Ultimate Conservative Difference Scheme V. A Second Order Sequel to Godunov's Method. *Journal of Computational Physics*, 32:101–136.

Van Leer, B. (1982) Flux Vector Splitting for the Euler Equations. In *Lecture Notes in Physics - Eighth International Conference on Numerical Methods in Fluid Dynamics*, pages 507–512, Berlin. Springer-Verlag. Volume 170.

Van Leer, B. (1985) On the Relation Between the Upwind-Differencing Schemes of Godunov, Enguist-Osher and Roe. *SIAM Journal on Scientific and Statistical Computing*, 5:1–20.

Van Leer, B. (1997) Godunov's Method for Gas Dynamics: Current Applications and Future Developments. *Journal of Computational Physics*, 132:1–2.

White, F. (1974) *Viscous Fluid Flow*. McGraw-Hill, New York.

Zalesak, S. (1979) Fully Multidimensional Flux-Corrected Transport Algorithms for Fluids. *Journal of Computational Physics*, 31:335–362.

Index

d in the United States
masters